9-4-85

To Sanjay,
  With appreciation
for your efforts
in our work on
surface forces.

                    Matt

INTERMOLECULAR AND SURFACE FORCES

With Applications to
Colloidal and Biological Systems

INTERMOLECULAR AND SURFACE FORCES

With Applications to
Colloidal and Biological Systems

JACOB N. ISRAELACHVILI

Department of Applied Mathematics
Institute of Advanced Studies
Australian National University
Canberra, Australia

1985

ACADEMIC PRESS
(Harcourt Brace Jovanovich, Publishers)
London   Orlando   San Diego   New York
Toronto   Montreal   Sydney   Tokyo

ACADEMIC PRESS INC. (LONDON) LTD.
24–28 Oval Road
LONDON NW1 7DX

United States Edition published by
ACADEMIC PRESS, INC.
Orlando, Florida 32887

British Library Cataloguing in Publication Data

Israelachvili, Jacob N.
    Intermolecular and surface forces.----
    1. Molecular theory   2. Surface chemistry
    I. Title   II. Series
    541.2'26   QD461
    ISBN 0-12-375180-2

Library of Congress Cataloging in Publication Data

Israelachvili, Jacob N.
    Intermolecular and surface forces.

    Bibliography: p.
    Includes index.
    1. Intermolecular forces. 2. Surface chemistry.
I. Title
QD461.I87   1985   541.2'26   84-21556
ISBN 0-12-375180-2 (alk. paper)

PRINTED IN THE UNITED STATES OF AMERICA

85 86 87 88   9 8 7 6 5 4 3 2 1

Contents

Part One
The Forces between Atoms and Molecules: Principles and Concepts

Chapter 1   Historical Perspective

Chapter 2   Some Thermodynamic Aspects of Intermolecular Forces

Chapter 3   Strong Intermolecular Forces: Covalent and Coulomb Interactions

Chapter 4   Interactions Involving Polar Molecules

Chapter 5   Interactions Involving the Polarization of Molecules

Chapter 6   van der Waals Forces

Chapter 7   Repulsive Forces, Total Intermolecular Pair Potentials, and Liquid Structure

Chapter 8   Special Interactions: Hydrogen Bonding, Hydrophobic, and Hydrophilic Interactions

Part Two
The Forces between Particles and Surfaces

Chapter 9   Some Unifying Concepts in Intermolecular and Interparticle Forces

Chapter 10   Contrasts between Intermolecular, Interparticle, and Intersurface Forces

Chapter 11   van der Waals Forces between Surfaces

Chapter 12   Electrostatic Forces between Surfaces in Liquids

Chapter 13 Solvation and Steric Interactions

Chapter 14 Adhesion

Part Three
Fluid-Like Structures:
Micelles, Bilayers, and Biological Membranes

Chapter 15 Thermodynamic Principles of Self-Association

Chapter 16   Aggregation of Amphiphilic Molecules into Micelles, Bilayers, Vesicles, and Biological Membranes

Chapter 17   The Interactions between Lipid Bilayers and Biological Membranes

Preface

Intermolecular forces embrace all forms of matter, and yet one finds very few university courses devoted to all aspects of this important subject. I wrote this book with the aim of presenting a comprehensive and unified introduction to intermolecular and surface forces, describing their role in determining the properties of simple systems such as gases, liquids, and solids but especially of more complex, and more interesting, systems. These are neither simple liquids nor solids, but rather a myriad of dissolved solute molecules, small molecular aggregates, or macroscopic particles interacting in liquid or vapour. It is the forces in such systems that ultimately determine the behaviour and properties of everyday things: soils, milk and cheese, paints and ink, adhesives and lubricants, many technological processes, detergents, micelles, biological molecules and membranes, and we ourselves—for each of us is one big biocolloidal system composed of about 75% water, as are most living organisms.

This subject therefore touches on a very broad area of phenomena in physics, chemistry, chemical engineering, and biology in which there have been tremendous advances in the past 15 years. These advances can be viewed in isolation within each discipline or within a broader multidisciplinary framework. The latter approach is adopted in this book, where I have tried to present a general view of intermolecular and surface forces with examples of the various and often seemingly disparate phenomena in which they play a role.

Because of the wide range of topics covered and the different disciplines to which the book is addressed, I have presumed only a basic knowledge of the "molecular sciences": physics (elementary concepts of energy and force, electrostatics), chemistry (basic thermodynamics and quantum mechanics), and mathematics (algebra and elementary calculus). The mathematical and theoretical developments, in particular, have been kept at a simple, unsophisticated level throughout. Vectors are omitted altogether. Most equations are derived from first principles followed by examples of how they apply to specific situations. More complicated equations are stated but are again carefully explained and demonstrated.

In a book such as this, of modest size yet covering such a wide spectrum, it has not been possible to treat each topic exhaustively or rigorously, and specialists may find their particular subject discussed somewhat superficially.

The text is divided into three parts, the first dealing with the interactions between atoms and molecules, the second with the interactions between "hard" particles and surfaces, and the third with "soft" molecular aggregates in solution such as micelles (aggregates of surfactant molecules) and biological membranes (aggregates of lipids and proteins). While the fundamental forces are, of course, the same in each of these categories, they manifest themselves in sufficiently different ways that, I believe, they are best treated in three parts.

The primary aim of the book is to provide a thorough grounding in the theories and concepts of intermolecular forces so that the reader will be able to appreciate which forces are important in whatever system he or she is dealing with and to apply these theories correctly to specific problems (research or otherwise). The book is intended for final-year undergraduate students, graduate students, and nonspecialist research workers.

I am deeply grateful to the following people who have read the text and made valuable comments for improving its presentation: Derek Chan, David Gruen, Bertil Halle, Roger Horn, Stjepan Marčelja, John Mitchell, Håkan Wennerström, and Lee White. My thanks also extend to Diana Wallace for typing the manuscript and to Tim Sawkins for his careful drawing of most of the figures. But above all, I am indebted to my wife, Karin, without whose constant support this book would not have been written.

Units and Symbols

Most of the published literature and equations on intermolecular and surface forces are based on the CGS system of units. In this book the *Système Internationale* (SI) is adopted. In this system the basic units are the kilogramme (kg) for mass, the metre (m) for length, and the second (s) for time; some old units such as gm (gramme = 10^{-3} kg) and cm (centimetre = 10^{-2} m) are still sometimes used although they are not part of the SI system. The basic electrical unit is the ampère (A). The SI system has many advantages over the CGS, not the least when it comes to forces. For example, force is expressed in newtons (N) without reference to the acceleration due to the earth's gravitation, which is implicit in some formulae based on the CGS system.

Derived SI units

Quantity	SI Unit	Symbol	Definition of Unit
Energy	Joule	J	$kg\,m^2\,s^{-2}$
Force	Newton	N	$kg\,m\,s^{-2} = J\,m^{-1}$
Pressure	Pascal	Pa	$N\,m^{-2}$
Electric charge	Coulomb	C	$A\,s$
Electric potential	Volt	V	$J\,A^{-1}\,s^{-1} = J\,C^{-1}$
Frequency	Hertz	Hz	s^{-1}

Fundamental constants

Constant	Symbol	SI	CGS
Avogadro's constant	N_0	$6.022 \times 10^{23}\,mol^{-1}$	$6.022 \times 10^{23}\,mole^{-1}$
Boltzmann's constant	k	$1.381 \times 10^{-23}\,J\,K^{-1}$	$1.381 \times 10^{-16}\,erg\,deg^{-1}$
Molar gas constant	$R = N_0 k$	$8.314\,J\,K^{-1}\,mol^{-1}$	$8.314 \times 10^7\,erg/mole\,deg$
Electronic charge	$-e$	$1.602 \times 10^{-19}\,C$	$4.803 \times 10^{-10}\,esu$
Planck's constant	h	$6.626 \times 10^{-34}\,J\,s$	$6.626 \times 10^{-27}\,erg\,sec$
Permittivity of free space	ε_0	$8.854 \times 10^{-12}\,C^2\,J^{-1}\,m^{-1}$	1
Rest mass of proton	m_p	$1.673 \times 10^{-27}\,kg$	$1.673 \times 10^{-24}\,gm$
Speed of light	c	$2.998 \times 10^8\,m\,s^{-1}$	$2.998 \times 10^{10}\,cm\,sec^{-1}$

Fraction	10^9	10^6	10^3	10^{-2}	10^{-3}	10^{-6}	10^{-9}	10^{-12}
Prefix symbol	G	M	k	c	m	μ	n	p

Conversion from CGS to SI

1 erg = 10^{-7} J
1 cal = 4.184 J
1 eV = 1.602×10^{-12} erg = 1.602×10^{-19} J
$1\,kT$ = 4.12×10^{-14} erg = 4.12×10^{-21} J at 298 K
1 kcal mole^{-1} = 4.184 kJ mol^{-1}
$1\,kT$ molecule^{-1} = 0.592 kcal mole^{-1} = 2.478 kJ mol^{-1} at 298 K
1 dyne = 10^{-5} N
1 dyne cm^{-1} = 1 erg cm^{-2} = 1 mN m^{-1} = 1 mJ m^{-2} (unit of surface tension)
1 dyne cm^{-2} = 10^{-1} Pa (N m^{-2})
1 atm = 1.013×10^{6} dyne cm^{-1} = 1.013×10^{5} Pa
1 Torr = 1 mm Hg = 133.3 Pa
1 esu = 3.336×10^{-10} C
1 Debye (D) = 10^{-18} esu cm = 3.336×10^{-30} C m (unit of electric dipole moment)

Conversion from SI to CGS

1 J = 10^{7} erg = 0.239 cal = 6.242×10^{18} eV
1 kJ mol^{-1} = 0.239 kcal mole^{-1}
1 N = 10^{5} dyne
1 Pa = 9.87×10^{-6} atm = 7.50×10^{-3} Torr = 10 dyne cm^{-2}
1 C = 3.00×10^{6} esu

Useful relations

kT/e = 25.69 mV at 298 K
1 C m^{-2} = 1 unit charge per 0.16 nm^2
κ^{-1} (Debye length) = $0.304/\sqrt{M}$ nm for 1:1 electrolyte at 298 K,
    where 1 M = 1 mol dm^{-3} $\equiv$ 6.022×10^{26} molecules per m^3

Common Symbols

A	Hamaker constant (J)	l, l_c	Length (m), critical hydrocarbon chain length (m)
a	Atomic or molecular radius (m), head-group area (m^2)	M	Concentration (mol dm^{-3})
C	Interaction constant, aqueous solute concentration in mole fraction units (mol dm^{-3}/55.5)	n	Refractive index
		P	Pressure (N m^{-2})
		Q, q	Charge (C)
D	Distance between two surfaces (m)	R	Radius (m)
E	Electric field strength (V m^{-1})	r	Distance between two atoms (m), radius (m)
F	Force (N)		
I	Ionization potential (J)	T	Temperature (K)
i	$\sqrt{-1}$	T_M, T_B	Melting or boiling points (K or °C)
k_c	Area compressibility modulus (J m^{-2})	T_c	Lipid-chain melting temperature
		u	Dipole moment (C m)
L	Latent heat (J mol^{-1})	U	Molar cohesive energy (J mol^{-1})

V, v Molar volume or volume (m^3)

W, w Interaction free energy (J), pair potential (J)

X Dimensionless concentration (e.g., mole fraction)

z Valency

α Polarizability $(C^2\ m^2\ J^{-1})$, interaction energy parameter

γ Surface tension–energy $(J\ m^{-2})$

ε Relative permittivity or dielectric constant

θ Angle

κ Inverse Debye length (m^{-1})

μ Chemical potential

μ^i, μ^0 Standard part of the chemical potential due to interactions

ν Frequency $(s^{-1}$ or Hz)

ρ Number density (m^{-3}) or mass density $(kg\ m^{-3})$

σ Atomic or molecular diameter (m), surface charge density $(C\ m^{-2})$, standard deviation

ϕ Angle

ψ Electrostatic potential (V)

ψ_0 Electrostatic surface potential (V)

$\approx$ approximately equal to

$\gtrsim$ slightly greater than

$\propto$ proportional to

$\sim$ very roughly equal to

$\geq$ greater than or equal to

Δ change or difference in

$>, <$ greater than, less than

$\gg$ very much greater than

$\langle\ \rangle$ average or mean

PART ONE

THE FORCES BETWEEN ATOMS AND MOLECULES: PRINCIPLES AND CONCEPTS

Chapter 1

Historical Perspective

1.1. The four forces of nature

It is now well established that there are four distinct forces in nature. Two of these are the *strong* and *weak interactions* that act between neutrons, protons, electrons, and other elementary particles. These two forces have a very short range of action, less than 10^{-5} nm, and belong to the domain of nuclear and high-energy physics. The other two forces are the *electromagnetic* and *gravitational interactions* that act between atoms and molecules (as well as between elementary particles). These forces are effective over a much larger range of distances, from subatomic to practically infinite distances, and are consequently the forces that govern the behaviour of everyday things (Fig. 1). For example, electromagnetic forces—the source of all intermolecular interactions—determine the properties of solids, liquids, and gases, the behaviour of particles in solution, chemical reactions, and the organization of biological structures. Gravitational forces account for tidal motion and many cosmological phenomena, and when acting together with intermolecular forces, determine such phenomena as the height that a liquid will rise in small capillaries and the maximum size that animals and trees can attain (Thompson, 1968).

This book is mainly concerned with intermolecular forces. Let us enter the subject by briefly reviewing its historical developments from the ancient Greeks to the present day.

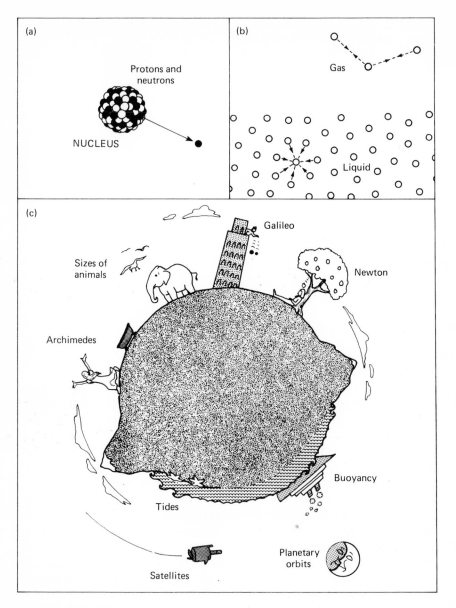

Fig. 1 The forces of nature.

(a)   Strong nuclear interactions hold protons and neutrons together in atomic nuclei. Weak interactions are involved in electron emission (β decay).

(b)   Electrostatic (intermolecular) forces determine the cohesive forces that hold molecules together in solids and liquids.

(c)   Gravitational forces affect tides, falling bodies, and satellites. Gravitational and intermolecular forces acting together determine the maximum possible sizes of mountains, trees, and animals.

1.2. Greek and medieval notions of intermolecular forces

The Greeks found that they needed only two fundamental forces to account for all natural phenomena. One was Love, the other was Hate. The first brought things together while the second caused them to part. The idea was first proposed by Empedocles around 450 B.C., was much "improved" by Aristotle, and formed the basis of chemical theory for 2000 years.

The ancients appear to have been particularly inspired by certain mysterious forces, or influences, that sometimes appeared between various forms of matter (forces that we would now label as magnetic or electrostatic). They were intrigued by the "action-at-a-distance" property displayed by these forces and were moved to reflect upon their virtues. What they lacked in concrete experimental facts they more than made up for by the abundant resources of their imagination. Thus, magnetic forces could cure diseases, though they could also cause melancholy and thievery; magnets could be used to find gold, and they were effective as love potions and for testing the chastity of women. Unfortunately, some magnetic substances lost their powers if rubbed with garlic (but they usually recovered when treated with goat's blood). Electric phenomena were endowed with attributes no less spectacular, manifesting themselves as visible sparks in addition to a miscellany of attractive influences that appeared when different bodies were rubbed together. All these wondrous practices, and much else, were enjoyed by our forebears until well into the seventeenth century.

1.3. The first scientific period: contrasts with gravitational forces

In the seventeenth century our subject entered its first phase of scientific scrutiny. Newton considered how the forces between molecules could be linked to the physical properties of matter, and later a number of eighteenth century researchers addressed themselves to the phenomenon of the capillary rise of liquids in glass tubes. In 1808 Clairaut suggested that capillarity could be explained if the attraction between the liquid and glass molecules was different from the attraction of the liquid molecules for themselves. It was also noticed that the height of rise of a liquid column does not depend on the capillary wall thickness, which led to the conclusion that these forces must be of very short range (or, in the language of the time, extended over "insensible" distances).

During the nineteenth century it was believed that one simple universal force law (similar to Newton's law for gravitational force) would eventually be found to account for all intermolecular attractions. To this end a number of interaction potentials were proposed that invariably contained the masses of the molecules, attesting to the belief at the time that these forces are related to gravitational forces. Thus, typical *interaction potentials* of two molecules were of the form $w(r) = -Cm_1m_2/r^n$, which is related to the *force law* $F(r)$ between them by

$$F(r) = -dw(r)/dr = -nCm_1m_2/r^{n+1}, \qquad (1.1)$$

where m_1, m_2 are the molecular masses, r their separation, C a constant, and n some integer believed to be 4 or 5, which may be compared with $n = 1$ for the gravitational interaction:

$$w(r) = -Gm_1m_2/r, \qquad G = 6.67 \times 10^{-11} \text{ N m}^2 \text{ kg}^{-2}. \qquad (1.2)$$

It is instructive to see how the power-law index n was so chosen. It arose from an appreciation of the fact that if intermolecular forces are not to extend over large distances, the value of n must be greater than 3. Why this is so can be simply established as follows: Suppose the attractive potential between two molecules or particles to be of the general form $w = -C/r^n$, where n is an integer. Now consider a region of space where the number density of these molecules is ρ. This region can be a solid, a liquid, a gas, or even a region in outer space extending over astronomical distances. Let us add all the interaction energies of one particular molecule with all the other molecules in the system. The number of molecules in a region of space between r and $(r + dr)$ away will be $\rho 4\pi r^2 \, dr$ (since $4\pi r^2 \, dr$ is the volume of a spherical shell of radius r and thickness dr, i.e., of area $4\pi r^2$ and thickness dr). The total interaction energy of one molecule with all the other molecules in the system will therefore be given by

$$\text{total energy} = -\int_\sigma^L w(r)\rho 4\pi r^2 \, dr = -4\pi C\rho \int_\sigma^L \frac{dr}{r^{n-2}}$$

$$= \frac{-4\pi C\rho}{(n-3)\sigma^{n-3}} \left[1 - \left(\frac{\sigma}{L}\right)^{n-3} \right] \qquad (1.3)$$

$$= -4\pi C\rho/(n-3)\sigma^{n-3} \qquad \text{for} \quad n > 3 \quad \text{and} \quad L \gg \sigma, \qquad (1.4)$$

where σ is the diameter of the molecules and L is the size of the system (e.g., the dimensions of a solid or the size of the box containing a gas). We can see that since σ must be smaller than L (i.e., $\sigma/L < 1$), large distance contributions to the interaction will disappear only for values of n greater than 3 (i.e., for $n = 4, 5, 6, \ldots$). But for n smaller than 3, the second term in Eq. (1.3) will be the greater, and the contribution from more distant molecules will dominate over that of nearby molecules. In such cases the size of the system must be taken into account, as occurs for gravitational forces where $n = 1$ and where distant planets, stars, and even galaxies are still strongly interacting with each other.

In later chapters we shall see that theoretical derivations of intermolecular force potentials do indeed predict that n always exceeds 3, and it is for this reason that the molecular properties of solids, liquids, and gases do not depend on the volume of material or on the size of the container (unless these are extremely small) but only on the forces between molecules in close proximity to each other. Important *long-range* intermolecular forces also exist, especially between macroscopic particles and surfaces, but their effective range of action rarely exceeds 100 nm.

Returning to the latter part of the nineteenth century, hopes for an all-embracing force law were rapidly dwindling as it became increasingly apparent that no suitable candidate would be forthcoming to explain the multitude of phenomena. However, by this time the modern concept of surface tension forces was firmly established, as was the recognition that these forces are the same as those that hold molecules together in solids and liquids and that in both cases they arise from interactions acting over very short distances (now known to be rarely in excess of a few molecular diameters). In addition, it was shown how these very short-range surface tension forces can account for such phenomena as capillarity, the shapes of macroscopic liquid droplets on surfaces, the contact angle between coalescing soap bubbles, and the breakup of a jet of water into spherical droplets (Fig. 2). Thus it was established that very short-range forces can lead to very long-range (i.e., macroscopic) effects. But further developments were to be forthcoming from quite different quarters: from work on gases rather than liquids.

1.4. First successful phenomenological theories

In an attempt to explain why real gases did not obey the ideal gas law ($PV = RT$, where P is the pressure, V the molar volume, R the gas constant, and T the temperature), the Dutch physicist J. D. van der Waals considered the effects of attractive forces between molecules (at a time when the very existence of molecules as we know them today was still being hotly debated). In 1873 he arrived at his famous equation of state for gases and liquids,

$$(P + a/V^2)(V - b) = RT, \qquad (1.5)$$

in which he subtracted the term b from the volume to account for the finite size of molecules and added the term a/V^2 to the pressure to account for the attractive intermolecular forces now known as *van der Waals forces*.

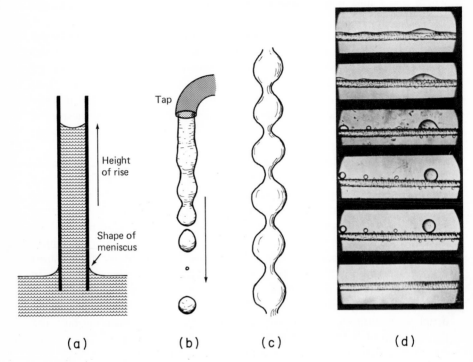

(a)   (b)   (c)   (d)

Fig. 2 Long-range effects produced by short-range forces.
(a) Capillary rise of liquids in narrow channels.
(b) Shape of water filament flowing out from a tap.
(c) Unduloid shape of a spider's web, a "frozen" intermediate state of (b).
(d) Action of detergent molecules in removing oily dirt from a fabric. Top to bottom: progressive addition of detergent diminishes the contact area of oily droplets on fibre until they finally detach. [(d) From Adam and Stevenson, 1953.]

By the early twentieth century it was recognized that intermolecular forces are not of a simple nature, and the pursuit of one basic force law now gave way to a less ambitious search for semiempirical expressions that could account for specific cases involving these forces. In this vein Mie, in 1903, proposed an interaction potential of the form

$$w(r) = -A/r^n + B/r^m, \tag{1.6}$$

which for the first time included a repulsive term as well as an attractive term. This was the first of a number of similar laws that successfully accounted for a wide range of phenomena, and it is still used today, as is the van der Waals equation of state. (Later we shall see how the parameters in potentials such as the Mie potential can be related to the constants a and b in the van der Waals equation of state.) Lamentably, it was soon found that many different potentials with a wide range of (adjustable) parameters would satisfactorily account for the same experimental data. Thus, while such empirical equations were useful, the nature and origin of the forces themselves remained a mystery.

1.5. Thermodynamic concepts make their debut

By the beginning of this century our subject had reached the end of its first scientific phase, coinciding (not unexpectedly) with the end of the classical era of physics and chemistry. But a number of important conceptual changes had also occurred. The period that had started with Newton and ended with van der Waals, Boltzmann, Maxwell, and Gibbs saw the abandonment of the purely mechanistic view of intermolecular forces and the adoption of thermodynamic and probabilistic concepts such as free energy and entropy. It was now appreciated that heat is also a form of energy and that the thermal energy and entropy associated with molecules are also involved in determining their behaviour (discussed in Chapter 2). In particular, it rapidly became apparent that there is a big gap between knowing what the forces between two isolated molecules are and understanding how an assembly of such molecules will behave. For example, the mere knowledge that air molecules attract each other does not mean that they will condense into a liquid or a solid at any given temperature or pressure; and even today there is no simple formula for deriving the properties of condensed phases from their intermolecular potentials, and vice versa.

1.6. Modern view of the origin
of intermolecular forces

Only with the elucidation of the electronic structure of atoms and molecules and the development of the quantum theory in the 1920s was it possible to understand the origin of intermolecular forces and derive expressions for their interaction potentials. It was soon established that all intermolecular forces are essentially electrostatic in origin. This is encapsulated in the Hellman–Feynman theorem, which states that once the distributions of the electron clouds have been determined from the solution of the Schrödinger equation, the intermolecular forces may be calculated on the basis of classical electrostatics. This theorem greatly simplified notions of the nature of intermolecular forces.

Thus, for two charges, we have the familiar inverse-square Coulomb force, while for moving charges we have electromagnetic forces and for the complex fluctuating charge distributions occurring in and around atoms, we obtain the various interatomic and intermolecular forces familiar in physics, chemistry, and biology.

This seems marvellously simple. Unfortunately, exact solutions of the Schrödinger equation are not easy to come by. In fact, it is even too difficult to solve (exactly) something as simple as two hydrogen atoms interacting in vacuum. For this reason, it has been found useful to classify intermolecular interactions into a number of seemingly different categories even though they all have the same fundamental origin.* Thus, such commonly encountered terms as ionic forces, van der Waals forces, hydrophobic, hydrogen bonding, and solvation forces are all a result of this classification, often accompanied by further divisions into strong and weak interactions and short-range and long-range forces. Such distinctions can be very useful, but as we shall see, they can also lead to confusion.

1.7. Recent trends: specialization and division

Today, as more and more information is accumulating on the properties of diverse systems at the molecular level, there is a natural desire to understand these phenomena in terms of the operative forces. Until recently there were three main areas of activity. The first was largely devoted to the forces acting between simple atoms and molecules in gases, where various quantum mechanical and statistical mechanical calculations are able to account for many of their physical properties (Hirschfelder *et al.*, 1954). The second area was concerned with the chemical binding of ions, atoms, and molecules in solids (Pauling, 1960), while the third dealt with the long-range interactions between small (colloidal) particles suspended in liquids (Verwey and Over-beek, 1948)—an area that is traditionally referred to as *colloid science.*

More recently the scope of endeavour has broadened to include liquid structure, surface phenomena, surfactant and polymer interactions, micelles, biological macromolecules, and biological membrane interactions (all of which will be considered in this book). The subject has now become so spread out that a tendency has developed for different disciplines to adopt their own terminology and concepts and even to emphasize quite different aspects of

---

* Some physicists believe that some of the four fundamental forces of nature may be likewise related.

interactions that are essentially the same. For example, in chemistry and biology emphasis is placed almost entirely on the *short-range* force fields around atoms and molecules, rarely extending more than one or two atomic distances. The language of the present-day molecular biologist is full of terms such as molecular packing, specific binding, lock and key mechanisms, and hydrogen bonds, all of which are essentially short range. In the different though closely related area of colloid science the emphasis is quite often on the *long-range* forces, which may determine whether two macromolecules, particles, or surfaces are able to get close enough in the first place before they can interact via the types of short-range forces mentioned above. In this discipline one is more likely to hear about electric double-layer forces, van der Waals forces, steric polymer interactions, etc., all of which are essentially long range.

The situation is not quite as silly as it sounds. Quite often the important interactions in chemistry and biology *are* the short-range ones, while those in colloid science are the long-range ones. But such contemporary issues no longer form part of our historical perspective. Their understanding requires some knowledge of the strength, range, and mode of action of different types of forces and thus leads us naturally into the subject matter of this book.

Chapter 2

Some Thermodynamic Aspects of Intermolecular Forces

2.1. Interaction energies of molecules in free space and in a medium

While this book is not primarily concerned with thermodynamics, it is nevertheless appropriate to start by considering some fundamental thermodynamic principles without which a mere knowledge of interaction forces will not always be very meaningful. In this chapter we shall introduce a number of simple but important thermodynamic relations and then illustrate how these, when taken together with the strengths of intermolecular forces, determine the properties of a system of molecules.

At the most basic molecular level we have the interaction potential $w(r)$ between two molecules or particles. This is usually known as the *pair potential* or, when the interaction takes place in a solvent medium, the *potential of mean force*. The interaction potential $w(r)$ is related to the force between two molecules or particles by $F = -dw(r)/dr$. Since the derivative of $w(r)$ with respect to distance r gives the force, and hence the work that can be done by the force, $w(r)$ is referred to as a *free energy* (a more appropriate term would be *available energy*).

In considering the forces between molecules in liquids, several effects are involved that do not occur when molecules interact in free space. Some of these effects are illustrated in Fig. 3 and will now be mentioned.

(i)   For two solute molecules in a solvent, their total pair potential $w(r)$ includes not only the direct solute–solute interaction energy but also any changes in the solute–solvent and solvent–solvent interaction energies as the two solute molecules approach each other. A dissolved solute molecule can

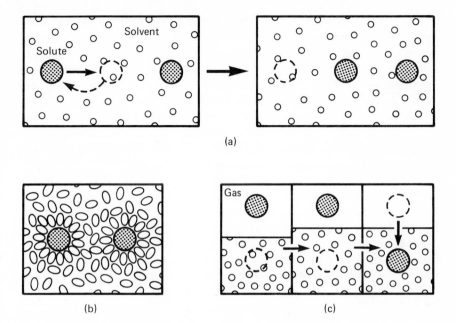

Fig. 3 Some solvent effects involved in the interactions of dissolved solute molecules or particles.
(a) Displacement of solvent by two approaching solute molecules.
(b) Solvation (reordering) of solvent molecules by solute.
(c) Cavity formation by solvent prior to solute insertion.

approach another only by displacing solvent molecules from its path (Fig. 3a). The net force therefore also depends on the attraction between the solute molecules and the solvent molecules. Thus, while two molecules may attract each other in free space, they may repel each other in a medium, for as two molecules move towards each other the work that must be done to displace the solvent may exceed that gained by the approaching solute molecules, and in such cases it becomes energetically more favourable for the two molecules to move apart.

(ii) Solute molecules often perturb the local ordering or structure of solvent molecules (Fig. 3b). If the free energy associated with this perturbation depends on the distance between two dissolved molecules, it produces an additional solvation force between them.

(iii) Solute–solvent interactions can change the properties of dissolved molecules, such as their dipole moments and charge (degree of ionization). The properties of dissolved molecules may therefore be different in different media.

(iv) Finally, when an individual molecule is introduced into a *condensed* medium, we must not forget the cavity energy expended by the medium when it forms the cavity to accommodate the guest molecule (Fig. 3c).

These effects are obviously interrelated and are collectively referred to as *solvent effects* or *medium effects*. They manifest themselves to different degrees depending on the nature and strength of solute–solute, solute–solvent, and solvent–solvent interactions. In later chapters we shall investigate the various functional forms of $w(r)$ for different types of forces and how solvent effects affect intermolecular and interparticle interactions. Meanwhile, let us address ourselves to the thermodynamic implications arising from the existence of interaction potentials without, at this stage, inquiring as to their origins.

When an individual molecule is in a medium (gas or liquid), it has what may be called a "cohesive energy" or a "self-energy" associated with it, given by the sum of its interactions with the surrounding molecules (including any change in the energy of the solvent molecules brought about by the introduction of the solute molecule). We shall denote this interaction free energy by μ^i. In many cases one needs to know the value of μ^i of an isolated molecule in a medium rather than its pair potential $w(r)$ with another individual molecule. For a molecule in the gas phase, we saw in Section 1.3 how one can estimate μ^i by summing all the potentials $w(r)$ over all of space. When $w(r)$ is a power law of the form

$$w(r) = -C/r^n, \qquad r > \sigma, \quad n > 3$$

$$= \infty, \qquad r < \sigma,$$

where σ is the so-called *hard sphere diameter* of the molecules, then for a molecule in the ideal gas phase, we may write

$$\mu^i_{gas} = \int_\sigma^\infty w(r)\rho 4\pi r^2 \, dr = -4\pi C\rho/(n-3)\sigma^{n-3}. \tag{2.1}$$

When a molecule is introduced from vapour into a condensed phase, μ^i must also include the cavity energy (Fig. 3c). For example, in a liquid or solid each molecule can have up to 12 other molecules in contact with it (known as *close packing*). If a molecule is introduced into its own liquid medium, then 12 liquid molecules must first separate from each other to form the hole. This costs $-6w(\sigma)$ in breaking the 6 "bonds" holding the 12 molecules together, where $w(\sigma)$ is the pair energy of two molecules in contact at $r = \sigma$. On introducing the guest molecule, 12 new "bonds" are formed costing $+12w(\sigma)$. The net energy change is therefore

$$\mu^i_{liq} \simeq 6w(\sigma), \tag{2.2}$$

which is *half* the total interaction energy of the molecule with its 12 nearest neighbours. Thus, the molar cohesive energy or latent heat of a simple liquid (or solid) may be expected to be

$$U = -N_0\mu^i_{liq} \approx -6N_0 w(\sigma). \tag{2.3}$$

It is interesting that a derivation similar to Eq. (2.1) also predicts a similar value for U. In a pure liquid (or solid) the number density of molecules ρ would be equal to

$$\rho \simeq 1/(\text{molecular volume}) \simeq 1/\{(4\pi/3)(\sigma/2)^3\},$$

where $\sigma/2$ is the molecular radius, and so we find that

$$\mu_{\text{liq}}^{i} = \frac{1}{2}\int_{\sigma}^{\infty} w(r)\rho 4\pi r^2\, dr \approx \frac{-12C}{(n-3)\sigma^n} \simeq \frac{12}{(n-3)}\, w(\sigma),$$

which for the van der Waals interaction where $n = 6$ (Chapter 6), gives

$$\mu_{\text{liq}}^{i} \approx 4w(\sigma) \qquad \text{or} \qquad U \approx -4N_0 w(\sigma).$$

As a rule of thumb we may therefore expect that the cohesive free energy of a molecule in a pure liquid or solid will be somewhere between 4 and 6 times the pair energy, the higher value being applicable to simple spherical molecules that condense into close-packed structures.

An *accurate* calculation of a molecule's free energy μ^i in a liquid from the pair potential $w(r)$ is extremely difficult. The mean number of molecules surrounding any particular molecule is not known in advance. It can be as high as 12 (for simple molecules such as the inert gases) and as low as 4 (for water). Further, the density ρ of neighbouring molecules is not uniform locally but rather it depends on the distance r from the molecule (cf. Fig. 3b). Thus, ρ in Eq. (2.1) should really be a function of r [i.e., $\rho(r)$]. This is known as the *density distribution*, which can be measured (see Chapters 7 and 13) but which can only be approximately determined a priori.

For a molecule dissolved in a solvent medium (i.e., when it is surrounded by *different* molecules), the calculation of its free energy in the medium becomes more difficult. Again, if we consider the simplest case of a solute molecule (s) surrounded by twelve solvent molecules (m) of similar size, then the change in cohesive energy on transferring the solute molecule from free space into the medium will be

$$\mu_{\text{liq}}^{i} \simeq -[6w_{\text{mm}}(\sigma) - 12w_{\text{sm}}(\sigma)]$$

since 6 solvent pairs must first be separated before the solute molecule can enter the medium and interact with the 12 solvent molecules. Note that if the solvent and solute molecules are the same, then $w_{\text{mm}}(\sigma) = w_{\text{sm}}(\sigma)$ and the above expression reduces to the earlier result, $\mu_{\text{liq}}^{i} \simeq 6w_{\text{mm}}(\sigma)$, as expected.

Finally, we may note that the effective pair potential between two dissolved solute molecules in a medium is just the change in their free energies μ^i as they approach each other.

2.2. The Boltzmann distribution

If the molecular interaction energy μ^i has different values μ^i_1 and μ^i_2 in two regions of a system (e.g., a liquid in equilibrium with its vapour), then at equilibrium the concentrations X_1 and X_2 of molecules in the two regions is given by the well-known *Boltzmann distribution*

$$X_1 = X_2 \exp\left(-\frac{\mu^i_1 - \mu^i_2}{kT}\right), \tag{2.4}$$

which may also be written as

$$\mu^i_1 + kT \log X_1 = \mu^i_2 + kT \log X_2, \tag{2.5}$$

where log means $\log_e$. Strictly, Eqs. (2.4) and (2.5) are exact only when molecules mix ideally in both regions, that is, for a dilute system. If there are many different regions or states in a system, each with different energies μ^i_n, then the condition of equilibrium is simply an extension of the above equation to all the states, viz.,

$$\mu^i_n + kT \log X_n = \text{constant} \quad \text{for all states } n = 1, 2, 3, \ldots$$
$$= \mu. \tag{2.6}$$

In other words, there will be a flow of molecules between all the different states of the system until Eq. (2.6) is satisfied (i.e., equilibrium is reached when the value of $\mu^i_n + kT \log X_n$ is uniform throughout). The quantity μ is known as the *chemical potential*, and it gives the *total* free energy per molecule (i.e., it includes the interaction energy as well as the contribution associated with its thermal energy). The $k \log X_n$ factor is known by a variety of names: the ideal gas entropy, the ideal solution entropy, the translational entropy, the entropy of dilution, and the entropy of mixing. The dimensionless concentrations X_n are usually expressed as mole fractions or volume fractions. For a pure solid or liquid, $X_1 = 1$ so that $\log X_1 = 0$.

2.3. The distribution of molecules and particles in systems at equilibrium

The requirement of equality of the chemical potentials, as expressed in Eq. (2.6), provides a very general and useful starting point for formulating conditions of equilibrium within a molecular framework (Castellan, 1972) and may be applied to a great variety of different systems [e.g., the equilibrium properties of small molecular aggregates such as surfactant micelles and lipid bilayers (Part III)]. In the remainder of this chapter we shall consider some simpler cases.

Suppose we wish to calculate how the number density ρ of molecules in the earth's atmosphere varies with altitude z. We may write

$$\mu_z^i + kT \log \rho_z = \mu_0^i + kT \log \rho_0,$$

so that

$$\rho_z = \rho_0 \exp\left(-\frac{\mu_z^i - \mu_0^i}{kT}\right).$$

Since $(\mu_z^i - \mu_0^i) = mgz$, where m is the molecular mass and g the gravitational acceleration, we immediately obtain

$$\rho_z = \rho_0 \exp(-mgz/kT),$$

which gives the density at height z in terms of the density at ground level ρ_0. This is the familiar gravitational or barometric distribution law.

In the above example the interaction energy did not arise from intermolecular interactions but from gravitational interactions. Let us now consider a two-state system where $\mu_1^i - \mu_2^i$ is the difference in energy due to the different intermolecular interactions in the two states or phases. If one of the states ($n = 1$) is a pure solid or liquid ($\log X_1 = 0$), we have

$$\mu_1^i = \mu_2^i + kT \log X_2,$$

thus

$$X_2 = \exp\left(-\frac{\mu_2^i - \mu_1^i}{kT}\right) = \exp\left(-\frac{\Delta\mu^i}{kT}\right). \tag{2.7}$$

Here, for example, μ_2^i could be the energy of molecules in solution relative to that in the solid (μ_1^i), whence X_2 is their *solubility*. Or μ_2^i could be the energy in a vapour in equilibrium with liquid; in this case we may write, using Eq. (2.1),

$$\mu_2^i = \mu_{\text{gas}}^i = -A\rho,$$

where $A = 4\pi C/(n-3)\sigma^{n-3} = $ constant, and

$$X_2 = \frac{1}{(v - B)} = \frac{\rho}{(1 - B\rho)}$$

for the effective density of the nonideal gas molecules, where v is the gaseous volume occupied per molecule and $B = 4\pi\sigma^3/3$ is the excluded volume since σ is the closest distance that one molecule can approach another. We therefore have for the chemical potential of a gas:

$$\mu = -A\rho + kT \log\left(\frac{\rho}{1 - B\rho}\right). \tag{2.8}$$

Now the pressure P is related to μ via (Moore, 1974; Hill, 1960)

$$(\partial\mu/\partial P)_T = v \qquad \text{or} \qquad (\partial P/\partial\rho)_T = \rho(\partial\mu/\partial\rho)_T \qquad (2.9)$$

since $\rho = 1/v$. Thus we find

$$P = \int_0^\rho \rho\left(\frac{\partial\mu}{\partial\rho}\right)_T d\rho = \int_0^\rho \left[-A\rho + \frac{kT}{(1-B\rho)}\right]d\rho$$

$$= -\frac{1}{2}A\rho^2 - \frac{kT}{B}\log(1-B\rho),$$

and for $B\rho < 1$, we can expand the log term as

$$\log(1-B\rho) = -B\rho - \tfrac{1}{2}(B\rho)^2 + \cdots$$

$$\approx -B\rho(1+\tfrac{1}{2}B\rho) \approx -B\rho/(1-\tfrac{1}{2}B\rho) \approx -B/(v-\tfrac{1}{2}B),$$

so that

$$P = -\frac{\tfrac{1}{2}A}{v^2} + \frac{kT}{(v-\tfrac{1}{2}B)}$$

or

$$(P + a/v^2)(v - b) = kT, \qquad (2.10)$$

which is the van der Waals equation of state in terms of the molecular parameters, $a = \tfrac{1}{2}A = 2\pi C/(n-3)\sigma^{n-3}$ and $b = \tfrac{1}{2}B = 2\pi\sigma^3/3$. In Chapter 6 we shall see how the constant a can be related to specific parameters describing the van der Waals intermolecular pair potential, though it should be noted that the van der Waals equation is neither rigorous nor exact but merely one of many equations that have been found useful for describing the PVT behaviour of gases.

2.4. The criterion of the thermal energy kT for gauging the strength of an interaction

As we have seen, the fundamental significance of the thermal energy kT has to do with the partitioning of molecules among the different energy levels of a system. However, the magnitude of kT is also often used as a rough indicator of the strength of an interaction, the idea being that if the interaction energy exceeds kT it will "win out" over the opposing or disorganising effects of thermal motion. However, molecular motion or disorder can appear in various ways (e.g., as translational or positional disorder or as orientational disorder), and it is therefore important to recognize how it manifests itself in different situations.

Let us first consider how strong intermolecular forces must be if they are to condense molecules into a liquid at a particular temperature and pressure. This amounts to finding how the interaction (or cohesive) energy μ^i is related to the boiling point of a substance. At standard atmospheric temperature and pressure (STP, where $T = 273$ K, $P = 1$ atm) one mole of gas occupies a volume of $\sim 22{,}400$ cm^3, while in the condensed state a typical value would be about 20 cm^3. Equating chemical potentials for gas and liquid molecules in equilibrium with each other gives

$$\mu^i_{\text{gas}} + kT \log X_{\text{gas}} = \mu^i_{\text{liq}} + kT \log X_{\text{liq}}, \tag{2.11}$$

and since μ^i_{liq} greatly exceeds μ^i_{gas} we may write

$$\mu^i_{\text{gas}} - \mu^i_{\text{liq}} \simeq -\mu^i_{\text{liq}} \simeq kT \log(X_{\text{liq}}/X_{\text{gas}}) \simeq kT \log(22{,}400/20)$$

$$\simeq 7kT.$$

If the gas obeys the gas law $PV \approx RT$, then at any other temperature the log term becomes $\log(22{,}400 \times T/20 \times 273)$. It is straightforward to verify that the log term is not very sensitive to temperature and that in the range $T = 100$ to 500 K the log term changes by only 13%. Over this range of temperature we therefore find

$$-\mu^i_{\text{liq}} \simeq 7kT_B \qquad \text{or} \qquad -\mu^i_{\text{liq}}/T_B \simeq 7k, \tag{2.12}$$

where T_B is the boiling point. This is an important result. First, it shows that the boiling point of a liquid is simply proportional to the energy needed to take a molecule from liquid into vapour. For one mole of molecules, since the energy of vapourization U_{vap} is given by

$$U_{\text{vap}} \simeq -N_0 \mu^i_{\text{liq}}, \tag{2.13}$$

we can write Eq. (2.12) *per mole* as

$$U_{\text{vap}}/T_B \simeq 7N_0 k \simeq 7R \simeq 60 \text{ J K}^{-1} \text{ mol}^{-1},$$

and we have derived, very crudely, the well-known relationship, known as *Trouton's rule*, which states that the latent heat of vapourization ($\simeq$ energy of vapourization) is related to the normal boiling point of a liquid (at 1 atm) by

$$U_{\text{vap}}/T_B \simeq 80 \text{ J K}^{-1} \text{ mol}^{-1}, \tag{2.14}$$

which corresponds to

$$-\mu^i_{\text{liq}} \simeq 10kT_B \quad \text{per molecule.} \tag{2.15}$$

Trouton's rule applies to a great variety of substances, as illustrated in Table I, and shows that the boiling point of a substance provides a reasonably accurate indication of the strength of the cohesive forces or energies holding molecules

TABLE I

Boiling points T_B and energies of vapourization U_{vap} of some common substances

Substance		T_B at 1 atm (K)	U_{vap} (kJ mol^{-1})	$U_{vap}/T_B{}^a$ (J K^{-1} mol^{-1})
Neon	Ne	27	1.8	65
Nitrogen	N_2	77	5.6	72
Argon	A	88	6.5	74
Oxygen	O_2	90	6.8	76
Methane	CH_4	112	8.2	73
Hydrogen chloride	HCl	188	16.2	86
Ammonia	NH_3	240	23.4	97
Hydrogen fluoride	HF	293	32.6	111
Ethanol	C_2H_5OH	352	39.4	112
Benzene	C_6H_6	353	30.8	87
Water	H_2O	373	40.7	109
Acetic acid	CH_3COOH	391	24.2	62
Iodine	I_2	456	41.7	91
Sodium	Na	1156	91.2	79
Lithium	Li	1645	129	78

^{a} For most "normal" substances, U_{vap}/T_B falls in the range of 70 to 90 J K^{-1} mol^{-1} (Trouton's rule). Higher values can usually be traced to cooperative association of molecules in the *liquid* (e.g., HF, NH_3, C_2H_5OH, and H_2O), while lower values to association, e.g., dimerization, in the *vapour*, as occurs for CH_3COOH (see Fig. 17d).

together in condensed phases. For solids, U_{vap} and T_B are replaced by the heat of sublimation and the sublimation temperature, respectively.

Second, Eq. (2.15) tells us that molecules will condense once their cohesive energy exceeds about $10kT$. Since we previously saw [Eq. (2.2)] that $\mu^i \simeq 6w(\sigma)$, we may further conclude that if the pair interaction energy of two molecules or particles in contact exceeds about $1.5kT$, then it is strong enough to condense them into a liquid or solid (see Table VIII, Chapter 6). It is for this reason that the thermal energy kT can be used as a standard reference for gauging the cohesive strength of an interaction potential, though it is essential to note that this indicator, and Trouton's rule, are valid only because of the particular value of the atmospheric pressure that exists on the earth's surface. This pressure determines that a gas molecule will occupy a volume of $v \simeq 4 \times 10^{-20}$ cm^3 at or near STP, which is needed for deriving Eqs. (2.12)–(2.15). It is also worth mentioning that the notion that molecules go into the vapour phase because of the kinetic energy ($\sim kT$) they acquire is not correct. Their kinetic energy does not disappear in a liquid or solid; the molecule's motion is merely restricted to a narrower region of space around a potential-energy minimum. The average translational kinetic energy of a molecule is $\frac{3}{2}kT$ irrespective of whether it is in the gas, liquid, or solid state. We shall not here consider the very complex energies associated with rotational and vibrational states of molecules in solids, liquids, and gases, which would take us outside the scope of this book.

So far the Boltzmann distribution has been used to find the *translational distribution* of molecules in different regions of a system. The Boltzmann distribution can also be used to determine the *orientational distribution* of molecules. For example, if the pair potential also depends on the mutual orientation of two anisotropic molecules [i.e., if $w(r)$ is also angle-dependent so that it may be written as $w(r,\theta)$], then the angular distribution of two molecules at a fixed distance r apart will be

$$X(\theta_2) = X(\theta_1)\exp\left\{-\frac{w(r,\theta_2) - w(r,\theta_1)}{kT}\right\}, \qquad (2.16)$$

and here again the factor kT appears as a convenient energy unit; but this time it appears for the strength of orientation-dependent interactions needed to mutually align molecules (e.g., solvent molecules) around a dissolved solute molecule.

2.5. Classification of forces

Intermolecular forces can be loosely classified into three categories. First, there are those that are *purely electrostatic* in origin arising from the *Coulomb force* between charges. The interactions between charges, permanent dipoles, quadrupoles, etc., fall into this category. Second, there are *polarization forces* that arise from the dipole moments *induced* in atoms and molecules by the electric fields of nearby charges and permanent dipoles. All interactions in a solvent medium involve polarization effects. Third, there are forces that are *quantum mechanical* in nature. Such forces give rise to *covalent bonding* (including charge-transfer interactions) and to the repulsive *exchange interactions* (due to the Pauli exclusion principle) that balance the attractive forces at very short distances.

These three categories should be considered as neither rigid nor exhaustive: for certain types of forces (e.g., van der Waals forces), an unambiguous classification is not possible, while some intermolecular interactions (e.g., magnetic forces) will not even be mentioned, since for the systems we shall consider, they are always very weak. A commonly encountered and even more

Type of interaction		Interaction energy $w(r)$
Covalent	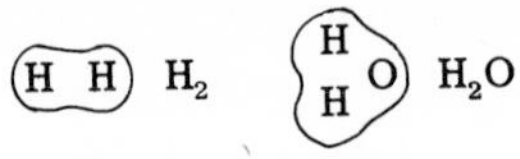 (H H) H_2 H_2O	Complicated, short range
Charge–charge	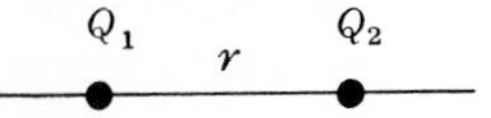 Q_1 ——r—— Q_2	$Q_1 Q_2/4\pi\varepsilon_0 r$ (Coulomb energy)
Charge–dipole	u θ ——r—— Q Fixed dipole	$-Qu\cos\theta/4\pi\varepsilon_0 r^2$
	u ——r—— Q Freely rotating	$-Q^2 u^2/6(4\pi\varepsilon_0)^2 kTr^4$
Dipole–dipole	u_1 θ_1 ——r—— ϕ u_2 θ_2 Fixed	$-u_1 u_2[2\cos\theta_1\cos\theta_2 - \sin\theta_1\sin\theta_2\cos\phi]/4\pi\varepsilon_0 r^3$
	u_1 ——r—— u_2 Freely rotating	$-u_1^2 u_2^2/3(4\pi\varepsilon_0)^2 kTr^6$ (Keesom energy)
Charge–nonpolar	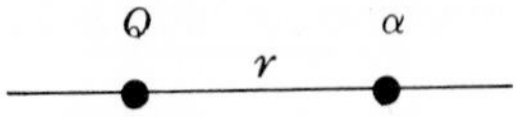 Q ——r—— α	$-Q^2\alpha/2(4\pi\varepsilon_0)^2 r^4$
Dipole–nonpolar	u θ r α Fixed	$-u^2\alpha(1 + 3\cos^2\theta)/2(4\pi\varepsilon_0)^2 r^6$
	u ——r—— α Rotating	$-u^2\alpha/(4\pi\varepsilon_0)^2 r^6$ (Debye energy)
Two nonpolar molecules	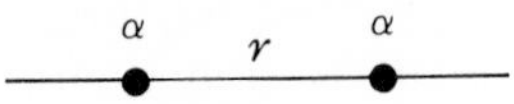 α ——r—— α	$-\left(\dfrac{3}{4}\right)\dfrac{h\nu\alpha^2}{(4\pi\varepsilon_0)^2 r^6}$ (London dispersion energy)
Hydrogen bond	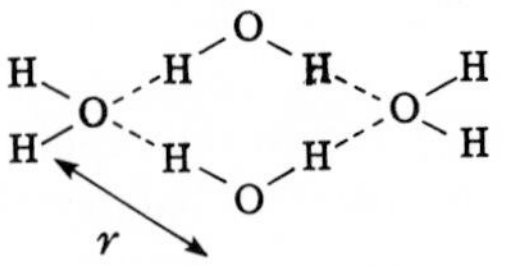 H, O, H—H, H diagram r	Complicated, short range, energy roughly proportional to $-1/r^2$

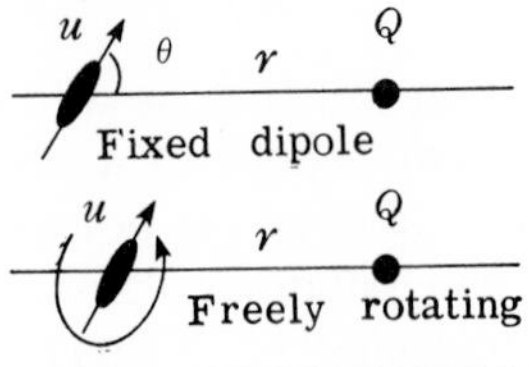

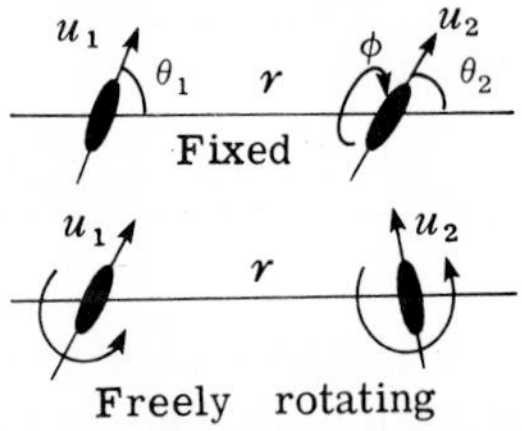

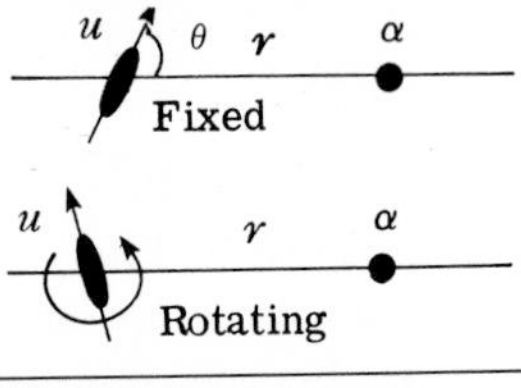

Fig. 4 Common types of interactions between atoms, ions, and molecules in vacuum. $w(r)$ is the interaction free energy (in J); Q, electric charge (C); u, electric dipole moment (C m); α, electric polarizibility (C^2 m^2 J^{-1}); r, distance between interacting atoms or molecules (m); k, Boltzmann constant (1.38×10^{-23} J K^{-1}); T, absolute temperature (K); h, Planck's constant (6.626×10^{-34} J s); ν, electronic absorption (ionization) frequency (s^{-1}); ε_0, permittivity of free space (8.854×10^{-12} C^2 J^{-1} m^{-1}). The force is obtained by differentiating the energy $w(r)$ with respect to distance r.

artificial classification of forces into short-range forces and long-range forces will not be adhered to here except in an intuitive sense whereby short-range forces refer to those interactions occurring at or very near atomic or molecular contacts.

Falling into the above categories are a number of fairly distinct interactions, as illustrated in Fig. 4. In the following sections and chapters these will be considered in turn, and in the process we shall introduce important conceptual aspects of intermolecular forces especially for interactions occurring in liquids.

Chapter 3

Strong Intermolecular Forces: Covalent and Coulomb Interactions

3.1. Covalent or chemical bonding forces

When two or more atoms come together to form a molecule, as when two hydrogen atoms and one oxygen atom combine to form a water molecule, the forces that tightly bind the atoms together within the molecule are called *covalent forces*, and the interatomic bonds formed are called *covalent bonds*. Closely allied to covalent bonds are metallic bonds. In both cases the bonds are characterized by the electrons being shared between two or more atoms so that the discrete nature of the atoms is lost.

Depending on the position an atom (or element) occupies in the periodic table, it can participate in a certain number of covalent bonds with other atoms. This number or stoichiometry is known as the atomic *valency*; for example, it is zero for the inert gases (e.g., argon), which cannot normally form covalent bonds with other atoms, one for hydrogen, two for oxygen, three for nitrogen, and four for carbon. A further characteristic of covalent bonds is their directionality, that is, they are directed or oriented at well-defined angles relative to each other. Thus, for multivalent atoms, their covalent bonds determine the way they will coordinate themselves in molecules or in crystalline solids to form an ordered three-dimensional lattice. For example, they determine the way carbon atoms arrange themselves to form the perfectly ordered diamond structure.

Covalent forces are of *short range*, that is, they operate over very short distances of the order of interatomic separations (0.1–0.2 nm). Table II shows the strength of some common covalent bonds. As can be seen they are mainly in the range 100–$300kT$ per bond (200–800 kJ mol^{-1}), and they tend to

TABLE II
Strengths of covalent bonds[a]

Bond Type		Strength (kJ mol^{-1})	Bond Type		Strength (kJ mol^{-1})
C≡N	(HCN)	870	Si—O		370
C=O	(HCHO)	690	C—C	(C_2H_6)	360
C=C	(C_2H_4)	600	C—O	(CH_3OH)	340
O—H	(H_2O)	460	N—O	(NH_2OH)	200
C—H	(CH_4)	430	F—F	(F_2)	150

[a] The strength of a covalent bond depends on the type of other bonds nearby in the molecule. For example, the C—H bond strength can be as low as 360 kJ mol^{-1} (in H—CHO) and as high as 500 kJ mol^{-1} (in HCN). Note that 1 kJ mol^{-1} corresponds to about $0.4kT$ per bond at 298 K.

decrease in strength with increasing bond length—a characteristic property of most intermolecular interactions.

3.2. Physical and chemical bonds

The complex quantum mechanical interactions that give rise to covalent bonding will not be a major concern in this book, which is devoted more to the interaction forces between unbonded *discrete* atoms and molecules; the latter are usually referred to as *physical forces* and give rise to *physical bonds*, in contrast to chemical forces, which give rise to *chemical* or *covalent bonds*.

Physical forces can be as strong as covalent forces, but they usually lack the specificity, stoichiometry, and strong directionality of covalent bonds. Indeed, the "bonds" they form should not really be considered as bonds at all, for during covalent binding the electron charge distributions of the uniting atoms change completely and merge; during physical binding they are merely perturbed, the atoms remaining as distinct entities. However, physical binding forces are strong enough to hold all but the smallest atoms and molecules together in solids and liquids at room temperature, as well as in colloidal and biological assemblies, and their long-range nature makes them the regulating forces in all phenomena that do not involve chemical reactions.

3.3. Coulomb forces or charge–charge interactions

The inverse-square Coulomb force between two charged atoms, or ions, is by far the strongest of the physical forces we shall be considering—stronger even than most chemical binding forces.

The free energy for the Coulomb interaction between two charges Q_1 and Q_2 is given by

$$w(r) = \frac{Q_1 Q_2}{4\pi\varepsilon_0\varepsilon r} = \frac{z_1 z_2 e^2}{4\pi\varepsilon_0\varepsilon r}, \tag{3.1}$$

where ε is the *relative permittivity* or *dielectric constant* of the medium and r the distance between the two charges. The expression on the right is commonly used for ionic interactions, where the magnitude and sign of each ionic charge is given in terms of the elementary charge ($e = 1.602 \times 10^{-19}$ C) multiplied by the ionic valency z. For example, $z = 1$ for monovalent *cations* such as Na^+, $z = -1$ for monovalent *anions* such as Cl^-, $z = 2$ for divalent cations such as Ca^{2+}, etc.

The Coulomb force F is given by

$$F = -\frac{dw(r)}{dr} = \frac{Q_1 Q_2}{4\pi\varepsilon_0\varepsilon r^2} = \frac{z_1 z_2 e^2}{4\pi\varepsilon_0\varepsilon r^2}. \tag{3.2}$$

For like charges, both w and F are positive and the force is repulsive, while for unlike charges they are negative and the force is attractive.

For completeness, we may also note that the electric field E at a distance r away from a charge Q_1 is defined by

$$E_1 = \frac{Q_1}{4\pi\varepsilon_0\varepsilon r^2} \quad \text{V m}^{-1}. \tag{3.3}$$

This field, when acting on a second charge Q_2 at r, gives rise to a force

$$F = Q_2 E_1 = \frac{Q_1 Q_2}{4\pi\varepsilon_0\varepsilon r^2},$$

which is the same as Eq. (3.2).

Let us put the strength of the Coulomb interaction into perspective. For two isolated ions (e.g., Na^+ and Cl^-) in contact, r is now the sum of the two ionic radii (0.276 nm), and the binding energy is

$$w(r) = -\frac{(1.602 \times 10^{-19})^2}{4\pi(8.854 \times 10^{-12})(0.276 \times 10^{-9})} = -8.4 \times 10^{-19} \quad \text{J}.$$

In terms of the thermal energy $kT = (1.38 \times 10^{-23})(300) = 4.1 \times 10^{-21}$ J at 300 K, this energy turns out to be of order $200kT$ per ion pair in vacuum. Only at a separation r greater than about 56 nm will the Coulomb energy fall below kT. We have thus established that the Coulomb interaction is very strong.

3.4. Ionic crystals

Coulomb forces (also known as *ionic forces*) hold the sodium and chloride ions together in the rigid salt lattice composed of alternate sodium and chloride ions, and the bond they give rise to is often referred to as the *ionic bond*. However, the above estimate of the binding energy of an isolated pair of Na^+–Cl^- ions does not apply to ions disposed within a crystal lattice; since Coulomb forces are of relatively long range [cf. the $1/r$ distance dependence of $w(r)$], the interaction energy of an ion with all the other ions in the lattice has to be summed, and not only with the nearest neighbours. Thus, in the NaCl crystal lattice each Na^+ has six nearest neighbour Cl^- ions at $r = 0.276$ nm, 12 Na^+ next nearest neighbours at $\sqrt{2}r$, eight Cl^- at $\sqrt{3}r$, etc.

The total interaction energy for a pair of Na^+–Cl^- ions in the lattice is therefore

$$
\begin{aligned}
\mu^i &= -\frac{e^2}{4\pi\varepsilon_0 r}\left[6 - \frac{12}{\sqrt{2}} + \frac{8}{\sqrt{3}} - \frac{6}{2} + \cdots\right] \\
&= -\frac{e^2}{4\pi\varepsilon_0 r}[6 - 8.485 + 4.619 - 3.000 + \cdots] \\
&= -1.748\,\frac{e^2}{4\pi\varepsilon_0 r} = -1.46 \times 10^{-18} \quad \text{J.}
\end{aligned}
\tag{3.4}
$$

The constant 1.748 is known as the *Madelung constant* and has different values for other crystal structures, varying between 1.638 and 1.763 for crystals composed of monovalent ions such as NaCl and CsCl, rising to about 5 for monovalent–divalent ion pairs such as CaF_2, and higher for multivalent ions such as SiO_2 and TiO_2. However, it is clear that the net (binding) energy is negative and that it is of the same order as for isolated ion pairs. To obtain the theoretical *molar lattice energy* or *cohesive energy* U of a NaCl crystal, we must multiply the above by Avogadro's number N_0, thus

$$
U = -N_0\mu^i = (6.02 \times 10^{23})(1.46 \times 10^{-18}) = 880 \quad \text{kJ mol}^{-1},
\tag{3.5}
$$

which is about 15% higher than the measured value due to our neglect of the repulsive forces at contact, which lower the final binding energy. These very short-range repulsive forces will be discussed in Chapter 7. The above value for the lattice energy of NaCl is fairly representative of other alkali halide lattice energies, which range from about 600 to 1000 kJ mol^{-1} going from RbI to LiF.

3.5. Reference states

It is important to always have the right reference state in mind when considering intermolecular interactions. Any value for the free energy is not very meaningful unless referred to some state with which it is being compared. Thus, when ions or molecules come together to form a condensed phase from the gaseous state, the reference state is at $r = \infty$, and the interaction occurs in vacuum ($\varepsilon = 1$). It is for this reason that Eq. (3.4) for the lattice energy of ionic crystals does not contain the dielectric constant of the medium, (e.g., $\varepsilon \approx 6$ for NaCl). On the other hand, if two ions are interacting in a condensed liquid medium, then the reference state is also at $r = \infty$, but the dielectric constant now appears in the denominator for the Coulomb interaction Eq. (3.1) since the interaction is now occurring entirely within the solvent medium.

3.6. Range of Coulomb forces

One very important aspect of Coulomb forces concerns the range of the interaction. The inverse-square distance dependence of the Coulomb force (the same as for gravitational forces) appears to make it very long ranged, in apparent contradiction with the statement made earlier that all intermolecular force laws must fall with distance faster than $1/r^4$ ($1/r^3$ for the energy). Since positive ions always have negative ions nearby, whether they are in a lattice or in solution, the electric field becomes *screened* and decays more rapidly away from them than from a truly isolated ion. In Chapter 12 we shall see that at large distances the decay is always exponential with distance, thus making all Coulomb interactions between ionic crystals, charged surfaces, and dissolved ions of much shorter range (though still of much longer range than covalent forces).

3.7. The Born energy of an ion

When a single ion is in a vacuum or in a medium, even though it may not be interacting with other ions, it still has an electrostatic free energy—equal to the electrostatic work done in forming the ion—associated with it. In vacuum this energy is referred to as the *self-energy*, while in a medium it is referred to as the *Born* or *solvation energy* of an ion, and it is an important quantity since it determines among other things the extent to which ions will dissociate or dissolve in different solvents. Let us see how the Born energy arises.

Imagine the process of charging an atom of radius a by gradually increasing its charge from zero up to its full charge Q. At any stage of this process let the ionic charge be q, and let this be incremented by dq. The work done in bringing this additional charge from infinity to $r = a$ is therefore, from Eq. (3.1) putting $Q_1 = q$, $Q_2 = dq$, and $r = a$,

$$dw = \frac{q\,dq}{4\pi\varepsilon_0\varepsilon a} \tag{3.6}$$

so that the total free energy of charging the ion, the Born energy, is

$$\mu^i = \int dw = \int_0^Q \frac{q\,dq}{4\pi\varepsilon_0\varepsilon a} = \frac{Q^2}{8\pi\varepsilon_0\varepsilon a} = \frac{(ze)^2}{8\pi\varepsilon_0\varepsilon a}. \tag{3.7}$$

The Born energy gives the electrostatic free energy of an ion in a medium of dielectric constant ε. In Chapter 5 we shall see how the Born energy can also be obtained from the pair potential for the interaction of an ion with a solvent molecule.

The change in free energy on transferring an ion from a medium of low dielectric constant ε_1 to one of high dielectric constant ε_2 is therefore negative, that is, energetically favourable, and equal to

$$\Delta\mu^i = -\frac{z^2e^2}{8\pi\varepsilon_0 a}\left[\frac{1}{\varepsilon_1} - \frac{1}{\varepsilon_2}\right] \quad J$$

$$= -\frac{28z^2}{a}\left[\frac{1}{\varepsilon_1} - \frac{1}{\varepsilon_2}\right] \quad kT \text{ per ion at } 300 \text{ K} \tag{3.8}$$

or

$$\Delta G = N_0\Delta\mu^i = -\frac{69z^2}{a}\left[\frac{1}{\varepsilon_1} - \frac{1}{\varepsilon_2}\right] \quad \text{kJ mol}^{-1}, \tag{3.9}$$

where a is given in nanometers. Thus, if one mole of monovalent cations and anions are transferred from the gas phase ($\varepsilon = 1$) into water ($\varepsilon = 78$), the gain in the molar free energy will be, assuming $a = 0.14$ nm for both cations and anions,

$$\Delta G = -\frac{2 \times 69}{0.14}\left[1 - \frac{1}{78}\right] \approx -1000 \quad \text{kJ mol}^{-1}. \tag{3.10}$$

Note that the Born energy does not include the energies expended by solvents in forming the cavities for accommodating ions; these are generally small compared to the large Born energy.

3.8. Solubility of ions in different solvents

Both the Coulomb energy and the Born energy are useful for understanding why ionic crystals such as Na^+Cl^-, in spite of their very high lattice energies, dissociate in water and in other solvents with high dielectric constants. If we consider the Coulomb interaction, Eq. (3.1), we immediately see that the electrostatic attraction between ions is much reduced (by a factor ε) in a medium. This is a somewhat superficial approach to the problem since there are serious difficulties in applying the Coulomb law at small interionic distances where the molecularity of the medium makes the continuum description (in terms of ε) break down. However, this approach does predict the right trends, so let us follow it up.

On the simplest level we may consider the difference in energy on going from the associated state to the dissociated state to be roughly given by Eq. (3.1). Thus, the free energy change in separating two monovalent ions such as Na^+ and Cl^- from contact in a solvent medium of dielectric constant ε is

$$\Delta\mu^i \approx \frac{+e^2}{4\pi\varepsilon_0\varepsilon}\frac{1}{(a_+ + a_-)},$$

where a_+ and a_- are the ionic radii of Na^+ and Cl^-. This energy is positive since the attractive Coulomb interaction will always favour association. However, some fraction of the ions will always dissociate due to their translational entropy or entropy of dilution (Chapter 2). The equilibrium concentration X_s of ions in a saturated solution will therefore be given by Eq. (2.7):

$$X_s = \exp(-\Delta\mu^i/kT) \approx \exp\left[-\frac{e^2}{4\pi\varepsilon_0\varepsilon kT}\frac{1}{(a_+ + a_-)}\right], \qquad (3.11)$$

where the value of the dimensionless parameter X_s may be identified with the solubility of an electrolyte in water, or in any solvent, in mole-fraction units. Thus, for NaCl in water, $a_+ + a_- = 0.276$ nm, $\varepsilon = 78$, so that at $T \approx 300$ K we expect very roughly

$$X_s \approx e^{-2.6} \approx 0.075,$$

which may be compared with the measured value of 0.10.

While Eq. (3.11) is far too simplistic to quantitatively account for the solubilities of electrolytes, it does predict the observed trends for monovalent salts reasonably well. For example, it predicts that the solubility of a salt in different solvents will be proportional to $e^{-\text{const}/\varepsilon}$, where ε is the solvent dielectric constant. Thus, a plot of $\log X_s$ against $1/\varepsilon$ should yield a straight line. This is more or less borne out in practice, as shown in Fig. 5 for NaCl and the amino acid glycine in different solvents. The large solubilizing power of water to ions is therefore seen as arising quite simply from its high dielectric constant (Table III) and not because of some special property of water.

From Eq. (3.11) we may also expect larger ions to be more soluble than smaller ions. This too is usually borne out in practice: alkali halide salts with large ionic radii, such as CsBr and KI, are generally much more soluble in various solvents than salts such as NaF and LiF (the latter has the smallest interionic distance and is the least soluble of the alkali halides in water).

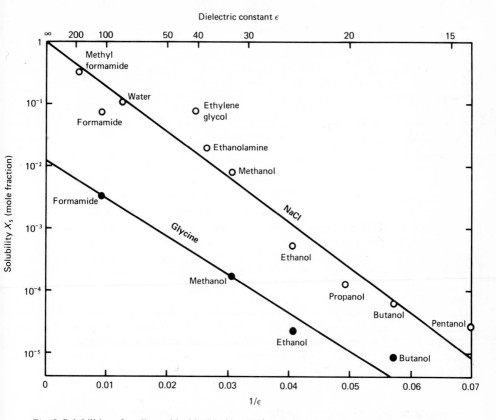

Fig. 5 Solubilities of sodium chloride ($NaCl \rightleftharpoons Na^+ + Cl^-$) and glycine ($NH_2CH_2COOH \rightleftharpoons NH_3^+CH_2COO^-$) in solvents of different dielectric constants ε at 25°C. Solubility (in mole fraction units) is plotted as log X_s as a function of $1/\varepsilon$. For NaCl, the line passes through $X_s = 1$ at $\varepsilon = \infty$, which from Eq. (3.11) suggests that the interaction of $Na^+ + Cl^-$ with these solvents is purely Coulombic. The slope of the line corresponds to an interionic spacing of 0.34 nm, about 0.06 nm larger than in the crystal. For glycine, the line tends to a finite value ($X_s < 1$) as ε tends to infinity, indicative of some additional type of solute–solute attraction—the van der Waals interaction (Chapter 6). Note that all the solvents are hydrogen-bonding liquids (Chapter 8). Non–hydrogen-bonding liquids are less effective as solvents for ionic species; for example, the solubility of NaCl in acetone ($\varepsilon = 20.7$) is $X_s = 4 \times 10^{-7}$ while that of glycine in acetone is $X_s = 2 \times 10^{-6}$. Solubility data were taken from *GMELINS Handbuch*, Series 21, Vol. 7 for NaCl, and from the *CRC Handbook of Chemistry and Physics* for glycine.

Note that the solubility of an electrolyte is strictly determined by the difference in the free energy of the ions in the lattice from that in solution. In the above approach we did not consider the lattice energy specifically; but it is also possible to approach the problem in a way that takes the lattice energy

TABLE III

Dielectric constants[a] ε of some common liquids and solids at 25°C

Compound		Dielectric Constant	Compound		Dielectric Constant
Hydrogen-bonding			Non–hydrogen bonding		
methyl-			Acetone	$(CH_3)_2CO$	20.7
formamide	$HCONHCH_3$	182.4	Chloroform	$CHCl_3$	5.0
Formamide	$HCONH_2$	109.5	Benzene	C_6H_6	⋅ 2.3
Hydrogen			Carbon		
fluoride	HF	84[b]	tetrachloride	CCl_4	2.2
Water	H_2O	78.5	Cyclohexane	C_6H_{12}	2.0
Formic acid	HCOOH	58.5[c]	Dodecane	$C_{12}H_{26}$	2.0
Ethylene			Hexane	C_6H_{14}	1.9
glycol	$C_2H_4(OH)_2$	40.7	Crystals		
Methanol	CH_3OH	32.6	Sodium		
Ethanol	C_2H_5OH	24.3	chloride	NaCl	6.0
n-Propanol	C_3H_7OH	20.2	Diamond	C	5.7
Ammonia	NH_3	16.9	Quartz	SiO_2	4.5
Acetic acid	CH_3COOH	6.2			

[a] The dielectric constant is a measure of the extent of reduction of electric fields and consequently of the reduced strengths of electrostatic interactions in a medium.
[b] Value at 0°C.
[c] Value at 16°C.

into account, by splitting up the dissociation process into two well-defined stages. The first is the dissociation of the crystal lattice into isolated gaseous ions, and the second is the transfer of these ions into the solvent. In this approach (see, e.g., Dasent, 1970; Pass, 1973) the energy associated with the first stage is simply the positive lattice energy, while the second stage reduces this by the negative Born energies of transferring the ions from the gas phase ($\varepsilon = 1$) into the solvent medium of dielectric constant ε. However, for water, the theoretical Born energies turn out to be much too large, even larger than the lattice energies [compare Eq. (3.5) with Eq. (3.10)], and so to obtain agreement with measured solubility and other thermodynamic data it has been found necessary to "correct" the crystal lattice radii of ions by increasing them by 0.02 to 0.09 nm when the ions are in water. The larger effective sizes of ions in water arises from their *solvation* or *hydration shells* (discussed in Chapter 4).

While both the Coulomb and Born energy approaches usually predict the right trends, neither are quantitatively reliable, because they ignore the complex and sometimes specific interactions that can occur between dissolved ions and the solvent molecules in their immediate vicinity. This problem has already been referred to as the *solvent effect*, and it is particularly acute when the solvent is water.

3.9. Solvent effects

The solvent effect is at the heart of all problems concerning the interactions between molecules (and surfaces) in a liquid medium at small separations, where the medium cannot always be treated as a structureless continuum defined solely in terms of its bulk properties such as its bulk dielectric constant, bulk density, etc., and where highly specific solvent effects can arise.

There are two basic approaches to this problem. The first is to assume that (at least some of) the bulk solvent properties such as the dielectric constant and density hold right down to molecular dimensions. In the second approach one starts by considering the specific interactions occurring at the atomic and molecular level and then attempts to build up the properties of the whole system. This dichotomy of approaches will be a recurring theme in theoretical treatments of various intermolecular forces. We shall refer to these two approaches as *continuum* and *molecular* theories. To understand the solvent effect for Coulomb interactions, we must investigate how a medium affects the electric fields around dissolved ions. Let us start with the continuum approach and attempt to assess its limitations by examining, this time in more detail, the origin of the Born and Coulomb energies.

3.10. Continuum approach

The Born energy of an ion is strictly a measure of the free energy associated with the electric field around the ion. From electrostatic theory (Guggenheim, 1949, Ch. 12; Landau and Lifshitz, 1963, Ch. II) the free energy density of an electric field arising from a fixed charge is $\frac{1}{2}\varepsilon_0\varepsilon E^2$ per unit volume, from which we may readily obtain the Born energy, Eq. (3.7), by integrating the energy density over all of space:

$$\mu^{i} = \frac{1}{2}\varepsilon_0\varepsilon \int E^2 \, dV = \frac{1}{2}\varepsilon_0\varepsilon \int_a^\infty \frac{Q^2}{(4\pi\varepsilon_0\varepsilon r^2)^2} 4\pi r^2 \, dr = \frac{Q^2}{8\pi\varepsilon_0\varepsilon a}. \tag{3.12}$$

Alternatively, if the dielectric constant of a medium changes, the variation in the free energy of a fixed charge is given by

$$\delta\mu^{i} = -\frac{1}{2}\varepsilon_0 E^2 \, \delta\varepsilon \quad \text{per unit volume.} \tag{3.13}$$

Again, by integrating the energy density over all of space we obtain

$$\delta\mu^i = -\frac{1}{2}\varepsilon_0\,\delta\varepsilon \int E^2\,dV = -\frac{1}{2}\varepsilon_0\,\delta\varepsilon \int_a^\infty \frac{Q^2}{(4\pi\varepsilon_0\varepsilon r^2)^2}4\pi r^2\,dr$$

$$= -\frac{Q^2\,\delta\varepsilon}{8\pi\varepsilon_0\varepsilon^2 a}. \tag{3.14}$$

From this we immediately obtain the change in the free energy on changing ε from ε_1 to ε_2 (equivalent to transferring the charge from one medium of ε_1 to another of ε_2) as follows:

$$\Delta\mu^i = -\frac{Q^2}{8\pi\varepsilon_0 a}\int_{\varepsilon_1}^{\varepsilon_2}\frac{d\varepsilon}{\varepsilon^2} = -\frac{Q^2}{8\pi\varepsilon_0 a}\left[\frac{1}{\varepsilon_1} - \frac{1}{\varepsilon_2}\right], \tag{3.15}$$

which gives the Born energy of transfer, Eq. (3.8).

The above derivations provide us with some important insights: First, the electrostatic self-energy is seen not to be concentrated on the ion itself, but rather it is spread out over the whole of space around the ion; thus, if in the above we integrate from $r = a$ to $r = R$, we find that the energy contained within a finite sphere of radius R around the ion is

$$\frac{Q^2}{8\pi\varepsilon\varepsilon_0}\left[\frac{1}{a} - \frac{1}{R}\right]. \tag{3.16}$$

For example, for an ion of radius 0.1 nm, 50% of its energy will be contained within a sphere of radius 0.2 nm and 90% within a radius of 1.0 nm. Thus, if the Born energy equation is to be applicable, the value of the dielectric constant of the medium must already be equal to the bulk value at ~ 0.1 nm away from the ion, a distance that is smaller than even the smallest solvent molecules!

Second, it can be shown that the Coulomb interaction in a medium can likewise be derived from the change in the electric field energy, integrated over the whole of space, when two charges are brought together. The Coulomb interaction can therefore be considered as the change in the Born energies of two charges as they approach each other. This is conceptually important since it shows that the Coulomb interaction in a medium is not determined by the dielectric constant in the region *between* two charges but by its value in the region *surrounding* (as well as between) the charges. It is for this reason that the strength of the Coulomb interaction of two oppositely charged ions will be reduced in a solvent medium even if the ions still remain in contact, that is, even before there are any solvent molecules between them! This is yet another manifestation of the long-range nature of electrostatic interactions. And it clearly shows why the extremely strong ionic "bond" is so easily disrupted in a medium of high dielectric constant, such as water, in marked contrast to covalent bonds, which—though weaker—are not generally broken by a solvent.

We have now established how a solvent medium affects the electrostatic Born and Coulomb energies of ions. We have seen that it is the value of ε of the locally surrounding medium that is important and that if the standard expressions of the Born and Coulomb energies (and other interaction energies that depend on ε) are to remain generally valid, the bulk value of ε must be attained already within the first shell of surrounding solvent molecules. In Section 4.8 we shall see that for water near monovalent ions, this is more or less expected on theoretical grounds. Experimentally, too, this is often the case, and we may recall how the Coulomb and Born energies are able to predict, at least semiquantitatively, the solubilities of monovalent ionic salts in different solvents. But we shall also encounter numerous instances where such continuum theories break down at small intermolecular distances.

3.11. Molecular approach

In reality two solute molecules do not see themselves surrounded by a homogeneous liquid medium but rather by discrete molecules of a given size and shape, and, for water, possessing a large dipole moment. As already noted in Section 2.1 there are no general theories that can handle this difficult problem except for the simplest of molecules (e.g., inert spherical molecules such as argon) that do not interact strongly with each other. In recent years much effort has been invested in developing molecular theories of liquids and of solute–solvent, solute–solute, and surface–surface interactions in liquids. Another approach is to model the solute and solvent atoms in terms of their sizes, shapes, permanent charge distributions, etc., and then allow a computer to work out how an assembly of such molecules will behave when they are allowed to interact according to some interaction potential, such as the Mie potential, Eq. (1.6), for simple molecules or more complex potentials for more complex molecules (e.g., water). Such *computer simulations* or *computer experiments* (see Chapters 7 and 13) are increasingly providing valuable insights into the physical properties of pure liquids at the molecular level, though similar studies of solute–solvent and solute–solute interactions are still in their infancy. In later chapters we shall return to explore the results and implications of computer experiments after we have acquired a better overview of the various intermolecular potentials that are involved in the simulations.

Chapter 4

Interactions Involving Polar Molecules

4.1. What are polar molecules?

Most molecules carry no net charge, but many possess an *electric dipole*. For example, in the HCl molecule the chlorine atom tends to draw the hydrogen's electron towards itself, and this molecule therefore has a permanent dipole. Such molecules are called *polar* molecules. The dipoles of some molecules depend on their environment and can change substantially when they are transferred from one medium to another, especially when molecules become ionized in a solvent. For example, the amino acid molecule glycine contains an acidic group on one side and a basic group on the other. In water this molecule exists as a dipolar molecule in the following form:

Glycine   Glycine
                           in water

Such dipolar molecules in water are also referred to as *zwitterions*. Quite often the magnitude of the positive and negative charges on zwitterions are not the same, and these molecules possess a net charge in addition to a dipole. Such molecules are often referred to as *dipolar ions*. It should already be apparent that the interactions and the solvent effects of polar molecules can be very complex.

The *dipole moment* of a polar molecule is defined as

$$u = ql, \qquad (4.1)$$

where l is the distance between the two charges $+q$ and $-q$. Thus, for two electronic charges $q = \pm e$ separated by $l = 0.1$ nm, the dipole moment is $u = (1.602 \times 10^{-19})(10^{-10}) = 1.6 \times 10^{-29}$ C m $= 4.8$ D. The unit of dipole moment is the *Debye*, where 1 Debye $= 1$ D $= 3.336 \times 10^{-30}$ C m. Small polar molecules have moments of the order of 1 D, some of which are listed in Table IV. Permanent dipole moments only occur in asymmetric molecules and thus not in single atoms. For isolated molecules, they arise from the asymmetric displacements of electrons along the covalent bonds, and it is therefore not surprising that a characteristic dipole moment can be assigned to each type of covalent bond. Table IV also lists some of these *bond moments*, which lie parallel to the lines of each bond. These values are approximate but very useful for estimating the dipole moments of molecules and especially of parts of molecules by vectorial summation of their bond moments. For

TABLE IV

Dipole moments of molecules, bonds, and molecular groups
(in Debye units)[a]

Molecules

Alkanes	0[b]	NH_3	1.47
C_6H_6 (benzene)	0	CH_3OH, C_2H_5OH	1.69
CCl_4	0	CH_3COOH (acetic acid)	1.7
CO_2	0	H_2O	1.85
$CHCl_3$ (chloroform)	1.06	C_2H_4O (ethylene oxide)	1.9
HCl	1.08	CH_3COCH_3 (acetone)	2.85

Bond moments

$C\!-\!H^+$	0.4	$C\!-\!C$	0	$C^+\!-\!Cl$	1.5–1.7
$N\!-\!H^+$	1.31	$C\!=\!C$	0	$N^+\!-\!O$	0.3
$O\!-\!H^+$	1.51	$C^+\!-\!N$	0.22	$C^+\!=\!O$	2.3–2.7
$F\!-\!H^+$	1.94	$C^+\!-\!O$	0.74	$N^+\!=\!O$	2.0

Group moments

$C\!-\!^+OH$	1.65	$C\!-\!^+CH_3$	0.4	$C\!-\!^+COOH$	1.7
$C\!-\!^+NH_2$	1.2–1.5	$C^+\!-\!NO_2$	3.1–3.8	$C\!-\!^+OCH_3$	1.3

[a] Data compiled from Wesson (1948), Smyth (1955), and Davies (1965).
[b] Depends on conformation (e.g., cyclopropane has a dipole moment).

example, the dipole moment of gaseous H_2O, where the H—O—H angle is $\theta = 104.5°$, may be calculated from

$$u_{H_2O} = 2\cos(\tfrac{1}{2}\theta)\,u_{OH} = 2\cos(52.25°) \times 1.51 = 1.85 \quad \text{D.}$$

4.2. Dipole self-energy

A dipole possesses an electrostatic self-energy μ^i that is analogous to the Born self-energy of an ion. The dipole self-energy is quite simply the sum of the Born energies of the two charges $\pm q$ (at infinity) minus the Coulomb energy of bringing the two charges together to form the dipole. Let us estimate this for two ions of equal radius a brought into contact to form a hypothetical dipolar molecule of length equal to the sum of the two ionic radii, $l = 2a$. We therefore have

$$\mu^i = \frac{1}{4\pi\varepsilon_0\varepsilon}\left[\frac{q^2}{2a} + \frac{q^2}{2a} - \frac{q^2}{r}\right] = \frac{q^2}{8\pi\varepsilon_0\varepsilon a} \quad \text{or} \quad \frac{u^2}{4\pi\varepsilon_0\varepsilon l^3}, \qquad (4.2)$$

where $r = l = 2a$. The dipole self-energy is therefore seen to be of roughly the same magnitude as the Born energy of an individual ion, Eq. (3.7), and its dependence on the dielectric constant of the medium is also the same. We may therefore expect that the solubility of polar molecules in different solvents should likewise increase with their value of ε. While this is generally the case, as shown in Fig. 5 (Chapter 3) for glycine, Eq. (4.2) for polar molecules is somewhat model dependent and not as useful as the Born equation for ions because as we have already noted, the dipole moment u can vary from solvent to solvent (Davies, 1965), and the energy of forming the cavity in the solvent for receiving the polar molecule and of other solute–solvent interactions (Onsager, 1936) are not included in Eq. (4.2).

4.3. Ion–dipole interactions

The second type of electrostatic pair interaction we shall consider is that between a charged atom and a polar molecule. As an illustrative example we shall derive the interaction potential for this case from basic principles. Figure 6 shows a charge Q at a distance r from the center of a polar molecule of dipole moment u subtending an angle θ to the line joining the two molecules. If the length of the dipole is l, with charges $\pm q$ at each end, then the total interaction energy will be the sum of the Coulomb energies of Q with $-q$ at B and Q with $+q$ at C:

$$w(r) = -\frac{Qq}{4\pi\varepsilon_0\varepsilon}\left[\frac{1}{AB} - \frac{1}{AC}\right], \qquad (4.3)$$

where

$$AB = [(r - \tfrac{1}{2}l\cos\theta)^2 + (\tfrac{1}{2}l\sin\theta)^2]^{1/2},$$
$$AC = [(r + \tfrac{1}{2}l\cos\theta)^2 + (\tfrac{1}{2}l\sin\theta)^2]^{1/2}.$$

(4.4)

At separations r exceeding the dipole's length l these distances can be written approximately as

$$AB \approx r - \tfrac{1}{2}l\cos\theta,$$
$$AC \approx r + \tfrac{1}{2}l\cos\theta,$$

and the interaction energy in this limit becomes

$$w(r) = w(r,\theta) = -\frac{Qq}{4\pi\varepsilon_0\varepsilon}\left[\frac{1}{r - \tfrac{1}{2}l\cos\theta} - \frac{1}{r + \tfrac{1}{2}l\cos\theta}\right]$$
$$= -\frac{Qq}{4\pi\varepsilon_0\varepsilon}\left[\frac{l\cos\theta}{r^2 - \tfrac{1}{4}l^2\cos^2\theta}\right]$$
$$= -\frac{Qu\cos\theta}{4\pi\varepsilon_0\varepsilon r^2} = -\frac{(ze)u\cos\theta}{4\pi\varepsilon_0\varepsilon r^2}.$$

(4.5)

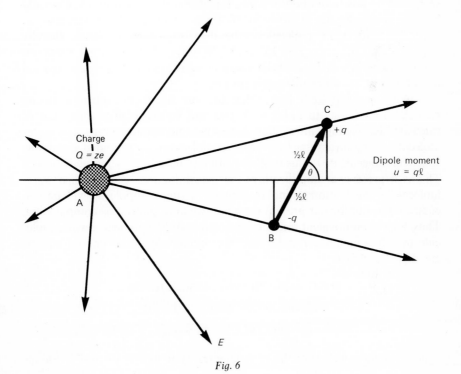

Fig. 6

Note that since the electric field of the charge acting on the dipole [Eq. (3.3)] is $E(r) = Q/4\pi\varepsilon_0\varepsilon r^2$, we see that in general the energy of a permanent dipole u in a field E may be written as

$$w(r, \theta) = -uE(r)\cos\theta. \tag{4.6}$$

Equation (4.5) gives the free energy for the interaction of a charge Q and a "point dipole" u (for which $l = 0$) in a medium. Thus, when a cation is near a dipolar molecule maximum attraction (i.e., maximum negative energy) will occur when the dipole points away from the ion ($\theta = 0°$), while if the dipole points towards the ion ($\theta = 180°$), the interaction energy is positive and the force is repulsive. Figure 7 shows how the energy varies with distance for a monovalent cation ($z = 1$) interacting with a dipolar molecule of moment 1 D in vacuum. The solid curves are based on the exact solution calculated from Eqs. (4.3) and (4.4), while the dashed curves are for the point-dipole approximation, Eq. (4.5), which shows itself to be surprisingly accurate down to fairly small separations. Only at ion–dipole separations r below about $2l$ does the approximate equation deviate noticeably ($> 10\%$) from that obtained using the exact formula. Thus, if the dipole moment arises from charges separated by less than about 0.1 nm, Eq. (4.5) will be valid at all physically realistic intermolecular separations. However, for greater dipole lengths—as occur in zwitterionic molecules—the deviations may be large, thereby requiring that the energy be calculated in terms of the separate Coulomb contributions. In such cases the interactions are always stronger than expected from Eq. (4.5), as can be inferred from Fig. 7.

It is also evident from Fig. 7 that the ion–dipole interaction is strong enough to bind ions to polar molecules and mutually align them. Let us calculate the vacuum interaction between some common ions and water molecules. We shall assume that the water molecule may be treated as a simple spherical molecule of radius 0.14 nm with a point dipole of moment 1.85 D. This is a gross oversimplification: the distribution of charges in a water molecule is much more complex than for a simple dipole as will be discussed later. But for our present purposes, we may ignore this complication. Thus, for the monovalent ion Na^+ ($z = 1$, $a = 0.095$ nm) near a water molecule ($a \approx 0.14$ nm, $u = 1.85$ D), the maximum interaction energy will be given by Eq. (4.5) as

$$w(r, \theta = 0°) = -\frac{(1.602 \times 10^{-19})(1.85 \times 3.336 \times 10^{-30})}{4\pi(8.854 \times 10^{-12})(0.235 \times 10^{-9})^2} = 1.6 \times 10^{-19} \quad J$$

$$= 39kT \quad \text{or} \quad 96 \text{ kJ mol}^{-1} \quad \text{at 300 K},$$

which compares well with the experimental value of 100 kJ mol^{-1} (Saluja,

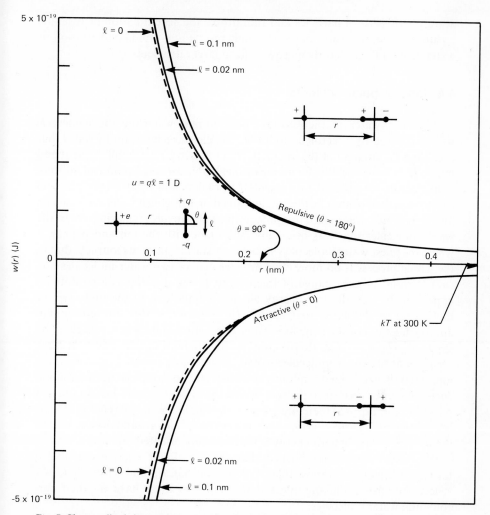

Fig. 7 Charge–dipole interaction energy in vacuum ($\varepsilon = 1$) between a unit charge e and a dipole of moment $u = ql = 1$ D oriented at different angles θ to the charge. Solid lines are exact solutions, Eqs. (4.3) and (4.4), for $l = 0.02$ nm and $l = 0.1$ nm; dashed lines are exact solutions for $l = 0$, which correspond to the approximate point-dipole formula, Eq. (4.5). Note the strength of the interaction relative to the thermal energy kT.

1976). For the smallest monovalent ion Li^+ ($a = 0.068$ nm), this rises to about $50kT$ or 125 kJ mol^{-1} (experimental value: 142 kJ mol^{-1}), while for the small divalent cations Mg^{2+} ($z = 2, a = 0.065$ nm) and Be^{2+} ($z = 2, a = 0.03$ nm), it will rise even further to about $100kT$ and $150kT$, respectively. The strongly attractive interaction between ions and water plays a role in the nucleation of atmospheric aerosol particles and rain drops; the accumulation of (usually) positive ions in the lower parts of clouds results in a rapid condensation of water around them that then grow to form larger droplets.

4.4. Ions in polar solvents

When ion–water interactions take place in bulk water the above energies will be reduced by a factor of about 80, the dielectric constant of water; but even then the strength of the interaction will exceed kT for small divalent and multivalent ions, and it is by no means negligible for small monovalent ions.

But before we proceed it is essential to understand what this interaction energy means. It cannot be the energy gained on bringing a water molecule up to an ion *in bulk water*, since this process must also involve replacing one of the water molecules that was originally in contact with the ion and placing it where the guest water molecule was before it started off on its journey, that is, the whole process is no more than an *exchange* of two water molecules and cannot result in any change of free energy. (Note that this is a continuum approach. In fact, the free energy change is finite due to the finite size of molecules, as discussed in Chapter 7.) However, within the continuum approach let us consider the ion–water interaction in bulk water as given by Eq. (4.5), where we note that it contains an orientation term $\cos \theta$. At large separations the water molecules would be randomly oriented relative to the ion, and if they remained randomly oriented right up to the ion, the interaction energy *would* be zero since the spatial average of $\cos \theta$ is zero. For an ion in a polar solvent, Eq. (4.5) therefore gives us an estimate of the free energy change brought about by *orienting* the polar solvent molecules around the ion, that is, the reference state of zero energy is for randomly oriented dipoles.

We have therefore established that the ion–dipole energies calculated for ions in water are comparable to or greater than kT and, from Eq. (2.16), reflect the strong aligning effect that small ions must have on their surrounding water molecules.

4.5. Strong ion–dipole interactions: hydrated ions

For small or multivalent ions in highly polar solvents, the strong orientation dependence of their ion–dipole interaction will tend to orient the solvent molecules around them, favouring $\theta = 0°$ near cations and $\theta = 180°$ near anions. Thus, in water Li^+, Be^{2+}, Mg^{2+}, and Al^{3+} have a number of water molecules orientationally bound to them. Such ions are called *solvated ions* or *hydrated ions*, and the number of water molecules they bind—usually between 4 and 6—is known as their *hydration number* (Table V). It should be noted,

TABLE V
Hydrated radii and hydration numbers of ions in water (approximate)

Ion	Bare Ion Radius (nm)	Hydrated Radius (nm)	Hydration Number (± 1)	Lifetime (Exchange Rate) (s)
H_3O^+	—	0.28	3	—
Li^+	0.068	0.38	5–6	10^{-9}–10^{-8}
Na^+	0.095	0.36	4–5	10^{-9}
K^+	0.133	0.33	3–4	10^{-9}
Cs^+	0.169	0.33	1–2	10^{-10}–10^{-9}
Be^{2+}	0.031	0.46	4^a	10^{-3}–10^{-2}
Mg^{2+}	0.065	0.43	6^a	10^{-6}–10^{-5}
Ca^{2+}	0.099	0.41	6	10^{-8}
Al^{3+}	0.050	0.48	6^a	10^{-1}–1
OH^-	0.176	0.30	3	
F^-	0.136	0.35	2	
Cl^-	0.181	0.33	1	
Br^-	0.195	0.33	1	
I^-	0.216	0.33	0	
NO_3^-	0.264	0.34	0	
$N(CH_3)_4^+$	0.347	0.37	0	

^{a} Number of water molecules forming a stoichiometric complex with the ion (e.g., $[Be(H_2O)_4]^{2+}$).

The hydration number gives the number of water molecules in the primary hydration shell, though the total number of water molecules affected is much larger and depends on the method of measurement. Similarly, the hydrated radius depends on how it is measured. Different methods can yield radii that can be as much as 0.1 nm smaller or larger than those shown. Table compiled from data given by Nightingale (1959), Amis (1975), Saluja (1976), Bockris and Reddy (1970), Cotton and Wilkinson (1980).

however, that these bound water molecules are not completely immobilized and that they do exchange with bulk water, albeit slowly. The hydration number is more of a qualitative indicator of the degree to which ions bind water rather than an exact value.

Closely related to the hydration number is the effective radius or *hydrated radius* of an ion in water, which is larger than its real radius (i.e., its crystal lattice radius) (Table V). Because smaller ions are more hydrated they tend to have larger hydrated radii than larger ions. Hydration numbers and radii can be deduced from measurements of the viscosity, diffusion, compressibility, conductivity, and various thermodynamic and spectroscopic properties of electrolyte solutions, the results rarely agreeing with one another (Amis, 1975; Saluja, 1976).

More insight into the nature of ion hydration can be gained by considering the time that water molecules remain bound to ions. In the pure liquid at room temperature the water molecules tumble about with a mean *reorientation time* or *rotational correlation time* of about 10^{-11} s. This also gives an estimate of the lifetime of the water–water bonds formed in liquid water (known as hydrogen bonds). But when water molecules are near ions various techniques, such as oxygen nuclear magnetic resonance, show that the mean lifetimes or exchange rates of water molecules in the first hydration shell can be much longer, varying from 10^{-11} s to many hours (Cotton and Wilkinson, 1980; Hertz, 1973). For very weakly solvated ions (usually large monovalent ions) such as $N(CH_3)_4^+$, Cl^-, Br^-, and I^-, these lifetimes are not much different from that for two water molecules, and they can even be shorter (referred to as *negative hydration*). For K^+, Na^+, and Li^+, the residence times of water molecules in the primary hydration shells are about 10^{-9} s. Divalent cations are always more strongly solvated than monovalent cations, and for Ca^{2+} and Mg^{2+}, the bound water lifetimes are about 10^{-8} s and 10^{-6} s, respectively. Even longer lifetimes are observed for very small divalent cations such as Be^{2+} (10^{-3} s), while for trivalent cations such as Al^{3+}, La^{3+}, and Cr^{3+} these can be many seconds or hours. In such cases the binding is so strong that an ion–water complex is actually formed of fixed stoichiometry (see Table V). In fact, these quasi-stable complexes begin to take on the appearance of (charged) molecules and are often designated as such (e.g., $[Mg(H_2O)_6]^{2+}$, $[Be(H_2O)_4]^{2+}$). In particular, protons H^+ always associate with one water molecule, which goes by the name of the *hydronium ion* or *oxonium ion* H_3O^+, while three water molecules are solvated around this ion to form $H_3O^+(H_2O)_3$. Likewise, the hydroxyl ion OH^- is believed to be solvated by three water molecules forming $OH^-(H_2O)_3$.

4.6. Solvation forces

The first shell of water molecules around a strongly solvated ion is usually referred to as the *primary hydration shell*. This is where the water molecules are most restricted in their motion. But the effect does not end there; it propagates out well beyond the first shell. Thus, in the second hydration shell the water molecules are freer to rotate and exchange with bulk water; in the third shell even more so, and so on.

Later we shall find that other types of interactions can also lead to a modified *solvent structure* or *molecular ordering* around solute molecules and surfaces and that the effect decays roughly exponentially with distance, extending a few molecular diameters. We shall refer to this modified solvent region as the *solvation zone* wherein the properties of the solvent (e.g., density, positional and orientational order, and mobility) are significantly different from the corresponding bulk values, as was shown schematically in Fig. 3b.

The existence of a solvation zone around dissolved ions, molecules, or particle surfaces in a solvent medium occurs whenever there are strong solute–solvent interactions (e.g., strong ion–dipole interactions in water) and has some important consequences. First, it affects the local dielectric constant of the solvent since the solvent molecules no longer respond to an electric field as they would in the pure liquid. The restricted mobility of water molecules around small ions would suggest that the effective dielectric constant should be lower in the solvation zone than in the bulk liquid. However, for some interactions, the reverse may occur. At present these effects are not well understood, but it is clear that solvation zones cannot be treated entirely in terms of continuum theories, since they arise from highly specific solute–solvent interactions occurring at the molecular level.

Second, when the solvation zones of two solvated molecules or surfaces overlap, a short-range force arises that again cannot be treated in terms of continuum models. For example, if the local dielectric constant of water around strongly solvated ions differs from 80, the short-range Coulomb interaction would be modified. But this is only one aspect of the problem. There has been much recent theoretical and experimental activity aimed at unravelling all the subtle effects associated with these solvent-mediated interactions that are now usually referred to as *solvation* or *structural forces* and, when water is the solvent, *hydration forces*. The nature of solvation forces will be investigated further in Chapters 7 and 13.

4.7. Dipole–dipole interactions

When two polar molecules are near each other there is a dipole–dipole interaction between them that is analogous to that between two magnets. For two point dipoles of moments u_1 and u_2 at a distance r apart and oriented relative to each other (as shown in Fig. 4), the interaction energy may be derived by a procedure similar to that used to obtain the energy for the charge–dipole interaction, and we find

$$w(r, \theta_1, \theta_2, \phi) = -\frac{u_1 u_2}{4\pi\varepsilon_0\varepsilon r^3}[2\cos\theta_1\cos\theta_2 - \sin\theta_1\sin\theta_2\cos\phi]. \qquad (4.7)$$

Maximum attraction therefore occurs when two dipoles are lying in line and is given by

$$w(r, 0, 0, \phi) = -2u_1 u_2/4\pi\varepsilon_0\varepsilon r^3. \qquad (4.8)$$

For two equal dipoles of moments 1 D, their interaction energy in vacuum will therefore equal kT at $r \approx 0.36$ nm when the dipoles are in line and at $r \approx 0.29$ nm when parallel. Since these distances are of the order of molecular separations in solids and liquids, we see that dipolar interactions are strong enough to bind only very polar molecules. Figure 8 shows the variation of the

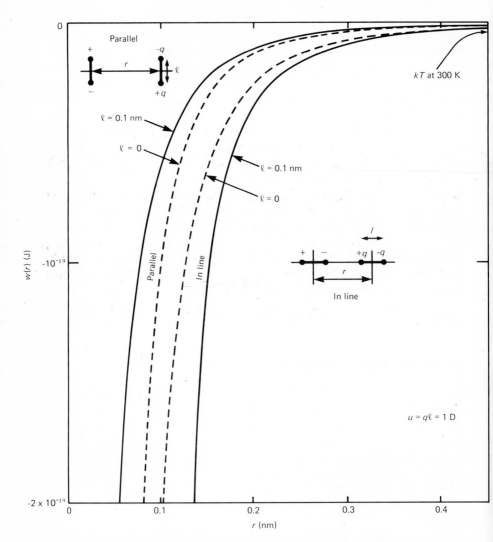

Fig. 8 Dipole–dipole interaction energy in vacuum between two dipoles each of moment 1 D. Note how much weaker this interaction is compared to the charge–dipole interaction and the large effect of finite dipole size.

pair interaction energy with distance for two point dipoles of moments 1 D approaching each other at these two orientations; the solid curves are the exact solutions for finite-sized dipoles, here assumed to be of length $l = 0.1$ nm, while the dashed curves are based on Eq. (4.7) for two point dipoles. In general we find that, even more than for the charge–dipole interaction, significant deviations from the ideal behaviour now occur for $r < 3l$, when Eq. (4.7) can no longer be used, and it is again necessary to analyse the interaction in terms of its individual charge–charge (Coulomb) contributions of which there will be four such terms for each pair of dipoles.

The dipole–dipole interaction is not as strong as the previous two electrostatic interactions we considered, and for dipole moments of order ~ 1 D, it is already weaker than kT at distances of about 0.35 nm in vacuum, while in a solvent medium this distance will be even smaller. This means that the dipole–dipole interaction (unlike the ion–dipole interaction) is not strong enough to lead to any strong mutual alignment of polar molecules in the liquid state. There are some exceptions, however, such as water whose very small size and large dipole moment does lead to short range association in the liquid. A glance at Table IV shows that the bond moments of O^-—H^+, N^-—H^+ and F^-—H^+ are unusually large. Since the electron-depleted H atom also has a particularly small size this means that other *electronegative* atoms such as —O, —N, and —F can get quite close to these highly polar X^-—H^+ groups and thus experience a very strong field. This results in a strong attractive force that can align neighbouring molecules possessing such groups (e.g., H_2O, NH_3, HF, and many others) in both the liquid and crystalline state. Such liquids are called *associated liquids*, and the special type of interaction they experience is known as the *hydrogen-bonding* interaction, which will be discussed further in Chapter 8. For the moment, let us investigate what happens when the orientation dependence of dipole–dipole and ion–dipole interaction energies are much *weaker* than the thermal energy kT.

4.8. Rotating dipoles and angle-averaged potentials

At large separations or in a medium of high ε, when the angle dependence of the interaction energy falls below the thermal energy kT, dipoles can now rotate more or less freely. However, even though the values of $\cos \theta$, $\sin \theta$, etc. when averaged over all of space are zero, the *angle-averaged potentials* are not zero since there is always a Boltzmann weighting factor that gives more weight to those orientations that have a lower (more negative) energy.

In general, the angle-averaged *free energy* $w(r)$ of an orientation-dependent interaction free energy $w(r, \Omega)$ is given by the important expression (see, e.g.,

Landau and Lifshitz, 1980, Ch. III; Rushbrooke, 1940)

$$e^{-w(r)/kT} = \int e^{-w(r,\Omega)/kT}\, d\Omega \Big/ \int d\Omega = \langle e^{-w(r,\Omega)/kT}\rangle, \tag{4.9}$$

where $d\Omega = \sin\theta\, d\theta\, d\phi$ corresponds to the polar and azimuthal angles θ and ϕ (see Fig. 4) and the integration is over all of the angular space. Accordingly, the denominator in Eq. (4.9) becomes

$$\int d\Omega = \int_0^\pi \sin\theta\, d\theta \int_0^{2\pi} d\phi = 4\pi,$$

so that in general we may write

$$e^{-w(r)/kT} = \langle e^{-w(r,\theta,\phi)/kT}\rangle = \frac{1}{4\pi}\int_0^\pi \int_0^{2\pi} e^{-w(r,\theta,\phi)/kT} \sin\theta\, d\theta\, d\phi. \tag{4.10}$$

We may note the spatially averaged values of some angles:

$$\langle \cos^2\theta\rangle = \frac{1}{4\pi}\int_0^\pi \cos^2\theta \sin\theta\, d\theta \int_0^{2\pi} d\phi = \frac{1}{3},$$

$$\langle \sin^2\theta\rangle = \tfrac{2}{3},$$

$$\langle \sin^2\phi\rangle = \langle \cos^2\phi\rangle = \tfrac{1}{2}, \tag{4.11}$$

$$\langle \sin\theta\rangle = \langle \cos\theta\rangle = \langle \sin\theta\cos\theta\rangle = 0,$$

$$\langle \sin\phi\rangle = \langle \cos\phi\rangle = \langle \sin\phi\cos\phi\rangle = 0.$$

When $w(r,\Omega)$ is *less* than kT we can expand Eq. (4.9):

$$e^{-w(r)/kT} = 1 - \frac{w(r)}{kT} + \cdots = \left\langle 1 - \frac{w(r,\Omega)}{kT} + \frac{1}{2}\left(\frac{w(r,\Omega)}{kT}\right)^2 - \cdots \right\rangle,$$

thus

$$w(r) = \left\langle w(r,\Omega) - \frac{w(r,\Omega)^2}{2kT} + \cdots \right\rangle. \tag{4.12}$$

The angle-averaged free energy for the *charge–dipole interaction* is therefore, using Eq. (4.5) for $w(r,\Omega)$,

$$w(r) = \left\langle -\frac{Qu\cos\theta}{4\pi\varepsilon_0\varepsilon r^2} - \left(\frac{Qu}{4\pi\varepsilon_0\varepsilon r^2}\right)^2 \frac{\cos^2\theta}{2kT} + \cdots \right\rangle$$

$$\approx -\frac{Q^2u^2}{6(4\pi\varepsilon_0\varepsilon)^2kTr^4} \quad \text{for} \quad kT > \frac{Qu}{4\pi\varepsilon_0\varepsilon r^2}, \tag{4.13}$$

which is attractive and very *temperature dependent*. Thus, for a monovalent ion interacting with the polar solvent molecules of a medium of dielectric constant

ε, Eq. (4.13) will supercede Eq. (4.5) at distances above $r \approx \sqrt{Qu/4\pi\varepsilon_0\varepsilon kT}$, which for a monovalent ion in water, setting $Q = e$, $u = 1.85$ D, $\varepsilon = 80$, becomes roughly 0.2 nm (i.e., about 0.1 nm out from an ion of radius 0.1 nm). We can now see why only water molecules of the first shell around ions sometimes become strongly restricted in their motion, and we may anticipate that this should be the sort of range around an ion over which the effective dielectric constant of the solvent may be substantially different from the bulk value! In the next chapter we shall see how Eq. (4.13) can also be used to derive the Born energy.

For the *dipole–dipole interaction*, a similar Boltzmann averaging of the interaction energy, Eq. (4.7), over all orientations $(\theta_1, \theta_2, \phi)$ leads to an angle-averaged interaction free energy of

$$w(r) = -\frac{u_1^2 u_2^2}{3(4\pi\varepsilon_0\varepsilon)^2 kTr^6} \quad \text{for} \quad kT > \frac{u_1 u_2}{4\pi\varepsilon_0\varepsilon r^3}. \tag{4.14}$$

The Boltzmann-averaged interaction between two permanent dipoles is usually referred to as the *orientation* or *Keesom* interaction. It is one of three important interactions (varying with the inverse sixth power of the distance) that contributes to the total *van der Waals* interaction between atoms and molecules; van der Waals forces will be discussed collectively in Chapter 6.

It is important to note that all the expressions so far derived give the *free energies* of interactions. From basic thermodynamics the Helmholtz *free energy* $\mathscr{F}$ of any system or interaction in related to the *internal energy* $\mathscr{E}$ by the well-known Gibbs–Helmholtz equation,

$$\mathscr{E} = \mathscr{F} + T\mathscr{S} = \mathscr{F} - T\left(\frac{\partial \mathscr{F}}{\partial T}\right)_V = -T^2\frac{\partial}{\partial T}\left(\frac{\mathscr{F}}{T}\right), \tag{4.15}$$

where $\mathscr{S}$ is the *entropy* of the system. Thus, for the angle-averaged dipole–dipole (Keesom) interaction in vacuum ($\varepsilon = 1$), we find

$$\mathscr{E} = -\frac{2u_1^2 u_2^2}{3(4\pi\varepsilon_0)^2 kTr^6}, \tag{4.16}$$

which is twice the Keesom *free* energy, Eq. (4.14), with which it is often confused in the literature. The distinction between $\mathscr{F}$ and $\mathscr{E}$ does not arise for temperature-independent pair interactions, since then $\mathscr{E} = \mathscr{F} - T(\partial \mathscr{F}/\partial T) = \mathscr{F}$.

4.9. Entropic effects

The reason why the interaction free energy (the energy available for doing work or the energy that gives the force) is less than the internal energy of two interacting dipoles is because some of the energy is taken up in aligning the dipoles as they approach each other, which ultimately transforms into heat. This is known as the entropic contribution to the interaction. Let us complete this chapter by considering these entropic effects a bit further.

From Eq. (4.15) the free energy may be written as

$$\mathscr{F} = \mathscr{E} - T\mathscr{S} = \mathscr{E} + T(\partial\mathscr{F}/\partial T) \qquad (4.17)$$

so that there is an entropic contribution $T(\partial\mathscr{F}/\partial T)$ that must be added to the total energy before we can know the free energy of an interaction. For example, for both the charge–dipole and dipole–dipole angle-averaged interactions *in vacuum*, we find that since $\mathscr{F} \propto 1/T$,

$$T(\partial\mathscr{F}/\partial T) = -\mathscr{F} \qquad (4.18)$$

so that

$$\mathscr{F} = \mathscr{E} - \mathscr{F} = \tfrac{1}{2}\mathscr{E}; \qquad (4.19)$$

thus, half the total energy is absorbed internally during the interaction. This energy is taken up in decreasing the rotational freedom of the dipoles as they become progressively more aligned on approach. Since $\mathscr{F}$ is negative, the entropic contribution $T(\partial\mathscr{F}/\partial T)$ to the free energy is *positive* (i.e., unfavourable), and since $T(\partial\mathscr{F}/\partial T) = -T\mathscr{S}$ we would say that the interaction is associated with a *loss of entropy*.

For interactions in a solvent medium, the medium dielectric constant generally appears in the denominator as ε or ε^2. Thus, if ε is temperature dependent, as occurs for polar solvents such as water (but not for nonpolar hydrocarbon solvents), the interaction energy now has an entropic component even if the interaction potential does not have an explicit temperature dependence. Thus, for the Born energy $\mathscr{F} = Q^2/8\pi\varepsilon_0\varepsilon a$, we find that

$$T\left(\frac{\partial\mathscr{F}}{\partial T}\right) = -\frac{Q^2}{8\pi\varepsilon_0\varepsilon a}\left(\frac{T}{\varepsilon}\frac{\partial\varepsilon}{\partial T}\right) = -\mathscr{F}\left(\frac{T}{\varepsilon}\frac{\partial\varepsilon}{\partial T}\right), \qquad (4.20)$$

and likewise for the Coulomb energy,

$$T\left(\frac{\partial\mathscr{F}}{\partial T}\right) = -\frac{Q_1Q_2}{4\pi\varepsilon_0\varepsilon r}\left(\frac{T}{\varepsilon}\frac{\partial\varepsilon}{\partial T}\right) = -\mathscr{F}\left(\frac{T}{\varepsilon}\frac{\partial\varepsilon}{\partial T}\right), \qquad (4.21)$$

while for the angle-averaged charge–dipole and dipole–dipole interactions in a medium we obtain

$$T\left(\frac{\partial\mathscr{F}}{\partial T}\right) = -\mathscr{F}\left(1 + \frac{2T}{\varepsilon}\frac{\partial\varepsilon}{\partial T}\right) = -T\mathscr{S}, \qquad (4.22)$$

where the first contribution to the entropy is associated with the orientational motion of the interacting *solute* dipoles as before [cf. Eq. (4.18)] and the second with the *solvent* molecules.

All the above entropic contributions that depend on $\partial \varepsilon / \partial T$ arise from changes in the orientational configurations of the *solvent* molecules associated with these interactions. For water,

$$\frac{1}{\varepsilon}\frac{\partial \varepsilon}{\partial T} = -0.00455 \quad \text{K}^{-1}, \tag{4.23}$$

which is negative. This allows us to make a number of predictions concerning the effects of interactions on the surrounding water molecules. Thus, for the Born interaction, $\mathscr{F}$ is positive so that $T(\partial \mathscr{F}/\partial T)$ is positive, and we may conclude that the solvation of ions by water is accompanied by a decrease in entropy, which indicates once again that water molecules become restricted in their translational and rotational freedom when they solvate ions. On the other hand, when two ions of opposite sign approach each other in a medium the entropy of the solvent increases.

The above considerations bring out the remarkable feature of the dielectric constant of a solvent in that it contains information on the entropic changes of the solvent molecules involved in an interaction; we shall make use of this property again when we consider other types of interactions.

Chapter 5

Interactions Involving
the Polarization of Molecules

5.1. The polarizability of atoms and molecules

We now enter the third and last category of electrostatic interactions we shall be considering—those that involve molecular *polarization*, that is, the dipole moments induced in molecules by the electric fields emanating from nearby molecules. Actually, we have already been much involved with polarization effects: whenever the macroscopic dielectric constant of a medium entered into our consideration this was no more than a reflection of the way the molecules of the medium are polarized by the local electric field. Here we shall look at these effects in more detail, starting at the molecular level. We shall find that apart from the purely Coulomb interaction between two charges or dipoles in vacuum, all the other interactions we have so far considered are essentially polarization-type interactions.

All atoms and molecules are polarizable. Their (dipole) *polarizability* α is defined according to the strength of the *induced* dipole moment u_{ind} they acquire in a field E, viz.,

$$u_{ind} = \alpha E. \tag{5.1}$$

For a nonpolar molecule, the polarizability arises from the displacement of its negatively charged electron cloud relative to the positively charged nucleus under the influence of an external electric field. For polar molecules, there are other contributions to the polarizability, discussed in the next section. For the moment, we shall concentrate on the polarizabilities of nonpolar molecules, which we shall denote by α_0.

52

As a simple illustrative example of how polarizability arises, let us imagine a one-electron atom whose electron of charge $-e$ circles the nucleus at a distance R; this would also define the radius of the atom. If under the influence of an external field E the electron orbit is shifted by a distance l from the nucleus (Fig. 9), then we have, for the induced dipole moment,

$$u_{ind} = \alpha_0 E = le. \tag{5.2}$$

Now the *external* force F_{ext} on the electron due to the field E is

$$F_{ext} = eE,$$

which must be balanced at equilibrium by the attractive force between the displaced electron orbit and the nucleus. This is none other than the Coulomb force $e^2/4\pi\varepsilon_0 R^2$ resolved along the direction of the field (Fig. 9). The *internal* (restoring) force is therefore

$$F_{int} = \frac{e^2}{4\pi\varepsilon_0 R^2}\sin\theta \approx \frac{e^2 l}{4\pi\varepsilon_0 R^3} \approx \frac{e}{4\pi\varepsilon_0 R^3}u_{ind}. \tag{5.3}$$

At equilibrium, $F_{ext} = F_{int}$, which leads to

$$u_{ind} = 4\pi\varepsilon_0 R^3 E = \alpha_0 E,$$

whence we obtain for the polarizability

$$\alpha_0 = 4\pi\varepsilon_0 R^3. \tag{5.4}$$

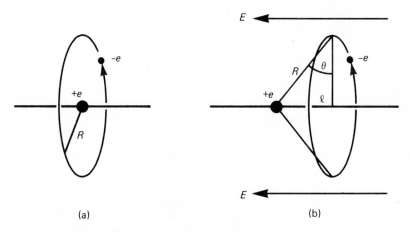

(a)   (b)

Fig. 9 Induced dipole in a one-electron atom.
(a)   No external field, $u = 0$.
(b)   In an external field E the electron's orbit is shifted by l from the positive nucleus so that the induced dipole moment is $u_{ind} = le = \alpha_0 E$, where the polarizability for this case is $\alpha_0 = 4\pi\varepsilon_0 R^3$.

The unit of polarizability is therefore $4\pi\varepsilon_0 \times$ (volume) or $C^2 \, m^2 \, J^{-1}$. The polarizability of atoms and molecules that arises from such electron displacements is known as *electronic polarizability*. Its magnitude, apart from the $4\pi\varepsilon_0$ term, is usually less than but of the order of the (radius)^3 of the atom or molecule. For example, for water, $\alpha_0/4\pi\varepsilon_0 = 1.48 \times 10^{-30} \, m^3 = (0.114 \, nm)^3$, where 0.114 nm is about 15% less than the radius of a water molecule (0.135 nm).

Table VI lists the electronic polarizabilities of some common atoms and molecules. Since the electronic polarizability is associated with displacements of electron clouds, it has long been recognised that the polarizability of a molecule can be obtained by simply summing the characteristic polarizabilities of its covalent *bonds*, since these are where the polarizable electrons are mostly localized. Table VI also lists some *bond polarizabilities*. As can be seen, the polarizability of methane, CH_4, is simply four times that of the C—H bond (i.e., $\alpha_{CH_4} = 4\alpha_{C-H}$). Likewise, the polarizability of ethylene, $CH_2{=}CH_2$, is given by $4\alpha_{C-H} + \alpha_{C=C}$. This additivity procedure is often

TABLE VI
Electronic polarizabilities α_0 of atoms,
molecules, bonds, and molecular groups[a]

Atoms and Molecules

He	0.20	NH_3	2.3	$CH_2{=}CH_2$	4.3
H_2	0.81	CH_4	2.6	C_2H_6	4.5
H_2O	1.48	HCl	2.6	Cl_2	4.6
O_2	1.60	CO_2	2.6	$CHCl_3$	8.2
Ar	1.63	CH_3OH	3.2	C_6H_6	10.3
CO	1.95	Xe	4.0	CCl_4	10.5

Bond Polarizabilities

Aliphatic	C—C	0.48	C=C	1.65
Aromatic	C⋯C	1.07	N—H	0.74
Aliphatic	C—H	0.65	C—Cl	2.60

Molecular Groups

| C—O—H | 1.28 | CH_2 | 1.84 |
| C—O—C | 1.13 | C=O | 1.36 |

[a] Polarizabilities α_0 are given in units of $(4\pi\varepsilon_0) \, Å^3 = (4\pi\varepsilon_0)10^{-30} \, m^3 = 1.11 \times 10^{-40} \, C^2 \, m^2 \, J^{-1}$. Note that when molecules are dissolved in a solvent medium their polarizability can change by up to 10%. Data compiled from Denbigh (1940), Hirschfelder *et al.* (1954), and Smyth (1955).

accurate to within a few percentage points, but it can fail for molecules in which the bonds are not independent of each other (delocalized electrons, as in benzene) or when nonbonded lone-pair electrons that also contribute to the polarizability are present, as in H_2O and other hydrogen-bonding groups. Under such circumstances it has been found useful to assign polarizabilities to certain molecular groups. Some *group polarizabilities* are also included in Table VI. For example, the polarizability of CH_3OH is $3\alpha_{C-H} + \alpha_{C-O-H} = 4\pi\varepsilon_0(3 \times 0.65 + 1.28) \times 10^{-30} = 4\pi\varepsilon_0(3.23 \times 10^{-30} \text{ m}^3)$.

5.2. The polarizability of polar molecules

In Section 5.1 we considered the polarizability arising solely from the electronic displacements in atoms and molecules. A freely rotating dipolar molecule (whose time-averaged dipole moment is zero) also has an *orientational polarizability*, arising from the effect of an external field on the Boltzmann-averaged orientations of the rotating dipole. Thus, in the presence of an electric field E these orientations will no longer time-average to zero but will be weighted along the field. If at any instant the permanent dipole u is at an angle θ to the field E, its resolved dipole moment along the field is $u\cos\theta$, and its energy in the field from Eq. (4.6) is $-uE\cos\theta$, so that the angle-averaged or induced dipole moment is given by (cf. Section 4.8)

$$u_{\text{ind}} = \langle u(\cos\theta)\exp(uE\cos\theta/kT)\rangle$$

$$= \frac{u^2E}{kT}\langle\cos^2\theta\rangle = \frac{u^2}{3kT}E, \qquad uE \ll kT. \tag{5.5}$$

Since u_{ind} is proportional to the field E, we see that the factor $u^2/3kT$ provides an additional contribution to the molecular polarizability. This is known as the *orientational* polarizability:

$$\alpha_{\text{orient}} = u^2/3kT. \tag{5.6}$$

The *total* polarizability of a polar molecule is therefore

$$\alpha = \alpha_0 + u^2/3kT \qquad \text{(the Debye–Langevin equation)}, \tag{5.7}$$

where u is its permanent dipole moment. Thus, for example, a polar molecule of moment $u = 1 \text{ D} = 3.336 \times 10^{-30}$ C m at 300 K will have an orientational polarizability of

$$\alpha_{\text{orient}} = \frac{(3.336 \times 10^{-30})^2}{3(1.38 \times 10^{-23})300} = 9 \times 10^{-40} \quad \text{C}^2 \text{ m}^2 \text{ J}^{-1}$$

$$= (4\pi\varepsilon_0)8 \times 10^{-30} \quad \text{m}^3,$$

a value that is comparable to the electronic polarizabilities α_0 of molecules. For a very polar molecule in a high field, or at a sufficiently low temperature, the molecule will become completely aligned along the field. When this happens (e.g., water near a small ion), the induced dipole moment is no longer proportional to E, and the whole concept of a molecule's "polarizability" breaks down.

5.3. Interactions between ions and uncharged molecules

When a molecule is at a distance r away from an ion, the electric field of the ion

$$E = (ze)/4\pi\varepsilon_0\varepsilon r^2 \tag{5.8}$$

will induce in the molecule a dipole moment of

$$u_{\text{ind}} = \alpha E = \alpha(ze)/4\pi\varepsilon_0\varepsilon r^2. \tag{5.9}$$

Thus, for a monovalent ion at a distance of 0.4 nm from a methane molecule in free space, the field on the molecule will be 9×10^9 V m^{-1}, and the induced dipole moment is

$$u_{\text{ind}} = \alpha_0 E = 4\pi\varepsilon_0(26 \times 10^{-30})(9 \times 10^9) = 2.6 \times 10^{-29} \quad \text{C m}$$

$$= 0.78 \quad \text{D.}$$

From Eq. (5.1) this corresponds to a unit charge separation in the molecule of $l \approx 0.016$ nm, which is about 4% of the molecular diameter.

Let us now consider the interaction between an ion and an uncharged molecule (Fig. 10). The induced dipole will point away from the ion (if a cation) and towards the ion (if an anion). In either case this will lead to an attractive force between the ion and the polarized molecule. The "reaction" field of the induced dipole acting back on the ion is, using Eq. (5.20),

$$E_r = -2u_{\text{ind}}/4\pi\varepsilon_0\varepsilon r^3 = -2\alpha E/4\pi\varepsilon_0\varepsilon r^3 = -2\alpha(ze)/(4\pi\varepsilon_0\varepsilon)^2 r^5, \tag{5.10}$$

so that the attractive force will be

$$F = -(ze)E_r = -2\alpha(ze)^2/(4\pi\varepsilon_0\varepsilon)^2 r^5, \tag{5.11}$$

which corresponds to an interaction free energy of

$$w(r) = -\alpha(ze)^2/2(4\pi\varepsilon_0\varepsilon)^2 r^4 \tag{5.12}$$

$$= -\tfrac{1}{2}\alpha E^2, \tag{5.13}$$

where E is the polarizing field of the ion acting on the molecule [Eq. (5.8)].

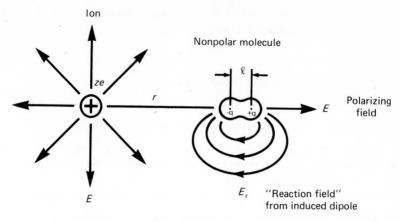

Fig. 10 A nonpolar molecule in a field E will acquire an induced dipole of moment $u_{ind} = ql = \alpha E$. The resulting force on the neutral molecule is therefore $F = q\,\Delta E$, where ΔE is the difference in E at either end of the dipole. Thus, $F = q(dE/dr)l = \alpha E(dE/dr)$, so that the interaction free energy is $w(r) = -\int_{\infty}^{r} F\,dr = -\frac{1}{2}\alpha E^2$. If the polarizing field E is due to an ionic charge, we obtain Eq. (5.12); if due to a dipole, we obtain Eq. (5.21).

It is important to note that this energy is half that expected for the interaction of an ion with an aligned dipole, which from Eq. (4.6) is

$$w(r) = -uE = -\alpha E^2. \tag{5.14}$$

The reason for this is that when a dipole moment is *induced* (rather than being *permanent* or *fixed*) some energy is taken up in polarizing the molecule. If we return to the example of Fig. 9, we see that this is the energy absorbed internally in displacing the positive and negative charges in the molecule and may be calculated by integrating the internal force F_{int} of Eq. (5.3) with respect to the charge separation from 0 to l, viz.,

$$w_{int}(r) = \int_{0}^{l} F_{int}\,dl = \int_{0}^{l} \frac{e^2}{4\pi\varepsilon_0 R^3}\,l\,dl = \frac{(el)^2}{8\pi\varepsilon_0 R^3} = \frac{(\alpha E)^2}{2\alpha}$$

$$= \frac{1}{2}\alpha E^2. \tag{5.15}$$

Thus, a factor $\frac{1}{2}\alpha E^2$ must be added to Eq. (5.14) to obtain the *total* free energy for the ion-induced dipole interaction [i.e., Eqs. (5.12)–(5.13)]. An alternative derivation of Eq. (5.13) is given in Fig. 10.

We may now insert the value for the total polarizability of polar molecules, $\alpha = \alpha_0 + u^2/3kT$, into Eq. (5.12) and obtain for the net ion-induced dipole interaction free energy

$$w(r) = -\frac{(ze)^2\alpha}{2(4\pi\varepsilon_0)^2 r^4} = -\frac{(ze)^2}{2(4\pi\varepsilon_0\varepsilon)^2 r^4}\left(\alpha_0 + \frac{u^2}{3kT}\right), \tag{5.16}$$

and we may note that the temperature-dependent term is identical to Eq. (4.13) derived earlier by a different method.

5.4. Ion–solvent molecule interactions and the Born energy

It is instructive to see how the ion-induced dipole interaction is related to the Born energy of an ion in a medium, which was previously discussed in Chapter 3. Let us first compute the total interaction energy of an ion in a medium with all the surrounding solvent molecules. For an ion of radius a, this can be calculated as before by integrating $w(r)$ of Eq. (5.12) or (5.16) over all of space:

$$\mu^i = \int_a^\infty w(r)\rho 4\pi r^2 \, dr = -\int_a^\infty \frac{\rho\alpha(ze)^2 4\pi r^2 \, dr}{2(4\pi\varepsilon_0\varepsilon)^2 r^4} = -\frac{\rho\alpha(ze)^2}{8\pi\varepsilon_0^2\varepsilon^2 a}, \quad (5.17)$$

where ρ is the number of solvent molecules per unit volume. To proceed further we have to relate the molecular polarizability α to the dielectric constant ε of the medium. This is a very complex problem and not well understood. However, the value of $(\rho\alpha)/\varepsilon_0$ in the above equation may in a first approximation be associated with the *electric susceptibility* χ of a medium. This is the polarizability per unit volume of a medium and is related to the dielectric constant ε by

$$\chi = (\varepsilon - 1). \quad (5.18)$$

The change in free energy $\Delta\mu^i$ when an ion goes from a medium of electric susceptibility χ_1 to one of χ_2 is therefore (since $d\chi = d\varepsilon$)

$$\Delta\mu^i = -\int_{\chi_1}^{\chi_2} \frac{(ze)^2}{8\pi\varepsilon_0\varepsilon^2 a} \, d\chi = -\int_{\varepsilon_1}^{\varepsilon_2} \frac{(ze)^2}{8\pi\varepsilon_0\varepsilon^2 a} \, d\varepsilon = -\frac{(ze)^2}{8\pi\varepsilon_0 a}\left[\frac{1}{\varepsilon_1} - \frac{1}{\varepsilon_2}\right], \quad (5.19)$$

which is the Born energy [Eq. (3.8)].

We have now seen how the Born energy—previously derived using a *continuum* analysis—can also be derived from a *molecular* approach. Furthermore, the molecular approach has the added advantage of providing insight into the limitations of the Born equation. Thus, for an ion in a polar solvent, we may expect the Born equation to be valid so long as the polarizabilities of the solvent molecules do not depend on their distance from the ion [i.e., so long as the polarizing field E and solvent dipole moment u are not so large that Eqs. (5.5)–(5.7) break down]; and we have already seen in Chapter 4 that for ions in a polar solvent, this condition is satisfied except for small or multivalent ions in very polar solvents (such as water).

5.5. Dipole-induced dipole interactions

The interaction between polar molecules and nonpolar molecules is analogous to the ion-induced dipole interaction just discussed except that the polarizing field comes from a permanent dipole rather than a charge. For a fixed dipole u oriented at an angle θ to the line joining it to a polarizable molecule (Fig. 4), since the magnitude of the electric field of a dipole is given by

$$E = \frac{u(1 + 3\cos^2\theta)^{1/2}}{4\pi\varepsilon_0\varepsilon r^3}, \tag{5.20}$$

the interaction energy is therefore

$$w(r, \theta) = -\frac{1}{2}\alpha_0 E^2 = -\frac{u^2\alpha_0(1 + 3\cos^2\theta)}{2(4\pi\varepsilon_0\varepsilon)^2 r^6}. \tag{5.21}$$

For typical values of u and α_0, the strength of this interaction is not sufficient to mutually orient the molecules, as occurs in ion–dipole or strong dipole–dipole interactions. The effective interaction is therefore the angle-averaged energy. Since the angle average of $\cos^2\theta$ is $\frac{1}{3}$, Eq. (5.21) becomes

$$w(r) = -\frac{u^2\alpha_0}{(4\pi\varepsilon_0\varepsilon)^2 r^6}, \tag{5.22}$$

while more generally, for two different molecules each possessing a permanent dipole moment u_1 and u_2 and polarizabilities α_{01} and α_{02}, their net dipole-induced dipole energy is

$$w(r) = -\frac{[u_1^2\alpha_{02} + u_2^2\alpha_{01}]}{(4\pi\varepsilon_0\varepsilon)^2 r^6}. \tag{5.23}$$

This is often referred to as the *Debye interaction* or the *induction interaction*. It constitutes the second of three inverse sixth power contributions to the total *van der Waals* interaction energy between molecules. The first we have already encountered in the angle-averaged dipole–dipole or Keesom interaction [Eq. (4.14)], which incidentally may also be obtained from Eq. (5.22) by replacing α_0 by $\alpha_{orient} = u^2/3kT$ so that for two dipoles u_1 and u_2, it gives the Keesom free energy:

$$w(r) = -\frac{u_1^2 u_2^2}{3(4\pi\varepsilon_0\varepsilon)^2 kTr^6}.$$

5.6. Unification of polarization interactions

Apart from the straight Coulomb interaction between two charges, all the other interactions we have considered have involved polarization effects, either explicitly for neutral molecules of polarizability α_0 or implicitly for rotating polar molecules that effectively behave as polarizable molecules of polarizability $\alpha = \alpha_0 + \alpha_{\text{orient}}$. For completeness, we may note that all these angle-averaged interactions may be expressed in one general equation; thus, for a charged polar molecule 1 interacting with a second polar molecule 2 we may write

$$w(r) = -\left(\frac{Q_1^2}{2r^4} + \frac{3kT\alpha_1}{r^6}\right)\frac{\alpha_2}{(4\pi\varepsilon_0\varepsilon)^2} \tag{5.24}$$

$$= -\left[\frac{Q_1^2}{2r^4} + \frac{3kT}{r^6}\left(\frac{u_1^2}{3kT} + \alpha_{01}\right)\right]\left(\frac{u_2^2}{3kT} + \alpha_{02}\right)\Big/(4\pi\varepsilon_0\varepsilon)^2, \tag{5.25}$$

where Q_1 is the charge of the first molecule and u_1, u_2, α_{01}, α_{02} are the dipole moments and electronic polarizabilities of the two molecules. Each of the six terms arising from the above equation may be identified (cf. Fig. 4) with a previous derivation (except for the small *van der Waals* $\alpha_{01}\alpha_{02}$ term discussed later). The unification of these various interactions is conceptually important, for it shows them all to be essentially polarization-type forces. If none of the molecules carries a net charge ($Q_1 = 0$), the above equation gives the Keesom-orientation and Debye-induction contributions to the total van der Waals force between molecules. The third contributor to van der Waals forces—the *dispersion force*—will be introduced in the next chapter.

Equation (5.25) is also useful for rapidly estimating the relative strengths of charge, dipole, and electronic polarizability contributions in an interaction. Thus, for typical values of $Q_1 = e = 1.6 \times 10^{-19}$ C, $u_1 = 1$ D $= 3.3 \times 10^{-30}$ C m, $\alpha_{01} = (4\pi\varepsilon_0)3 \times 10^{-30}$ m^3, and $r = 0.5$ nm, at $T = 300$ K the ratio of the three terms in the square brackets $(Q_1^2/2r^4):(u_1^2/r^6):(3kT\alpha_{01}/r^6)$ is about $800:3:1$. For water, however, because of its large dipole moment u and unusually small polarizability α_0, the ratio $u^2:3kT\alpha_0$ is about $20:1$ (i.e., the permanent-dipole-associated interactions of water always dominate over electronic polarization effects).

5.7. Solvent effects and "excess polarizabilities"

The interaction between molecules or small particles in a solvent medium can be very different from that of isolated molecules in free space (i.e., in a gas). The presence of a suspending solvent medium does more than simply reduce the interaction energy or force by a factor ε^2, as might appear from Eqs. (5.24)–(5.25). First, the effective dipole moment u of a molecule can be quite different in the liquid state or when dissolved in a medium and depends in a complicated way on its interactions with the surrounding solvent molecules. This will alter its effective polarizability α in a way that can only be found by experiment.

Second, as already discussed in Section 2.1, a dissolved molecule can move only by displacing an equal volume of solvent from its path; hence, the polarizability α in a medium must represent the *excess polarizability* of a molecule or particle over that of the solvent and must vanish when a dissolved particle has the same properties as the solvent. Qualitatively, we may say that if no electric field is reflected by a particle, it is "invisible" in the solvent medium and consequently does not experience a force.

The complicated problem of knowing the excess or effective polarizability of a molecule or particle can be approached by treating a molecule or a small particle as a dielectric medium of a given size and shape. This has an obvious advantage since the dielectric constant of a medium is easily measurable. Accordingly, a molecule i may be modelled as a dielectric sphere of radius a_i and dielectric constant ε_i. Now in a medium of dielectric constant ε (Fig. 11a), such a dielectric sphere will be polarized by a field E and acquire an excess dipole moment given by (Bleaney and Bleaney, 1959, p. 44; Landau and

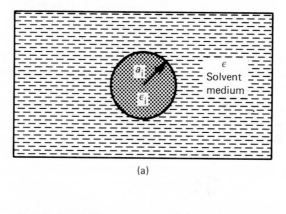

(a)

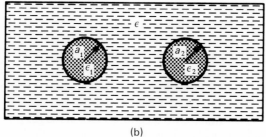

(b)

Fig. 11 A dissolved molecule or small solute particle can be modelled as a sphere of radius a_i and dielectric constant ε_i. Its polarizability in a medium of dielectric constant ε is $\alpha_i = 4\pi\varepsilon_0\varepsilon a_i^3(\varepsilon_i - \varepsilon)/(\varepsilon_i + 2\varepsilon)$.

Lifshitz, 1963, p. 44)

$$u_{\text{ind}} = 4\pi\varepsilon_0\varepsilon\left(\frac{\varepsilon_i - \varepsilon}{\varepsilon_i + 2\varepsilon}\right)a_i^3 E$$

so that its effective or *excess polarizability* in the medium is

$$\alpha_i = 4\pi\varepsilon_0\varepsilon\left(\frac{\varepsilon_i - \varepsilon}{\varepsilon_i + 2\varepsilon}\right)a_i^3 = 3\varepsilon_0\varepsilon\left(\frac{\varepsilon_i - \varepsilon}{\varepsilon_i + 2\varepsilon}\right)v_i, \qquad (5.26)$$

where $v_i = \frac{4}{3}\pi a_i^3$ is the volume of the molecule or sphere. This equation shows that for a dielectric sphere of high ε_i in free space ($\varepsilon = 1$), its polarizability is roughly $\alpha_i \approx 4\pi\varepsilon_0 a_i^3$, as previously found for a simple one-electron atom [Eq. (5.4)]. Further, if $\varepsilon > \varepsilon_i$, the polarizability is *negative*, implying that the direction of the induced dipole is opposite to that in free space.

If we substitute Eq. (5.26) into Eq. (5.24), we obtain the important result

$$w(r) = -\left[\frac{Q_1^2}{8\pi\varepsilon_0\varepsilon r^4} + \frac{3kT}{r^6}\left(\frac{\varepsilon_1 - \varepsilon}{\varepsilon_1 + 2\varepsilon}\right)a_1^3\right]\left(\frac{\varepsilon_2 - \varepsilon}{\varepsilon_2 + 2\varepsilon}\right)a_2^3, \qquad (5.27)$$

which allows us to conclude the following:

(i)   The net force between dissolved molecules or small particles in a medium (Fig. 11b) can be zero, attractive, or repulsive, depending on the relative magnitudes of ε_1, ε_2, and ε.

(ii)   Ions will be attracted to dissolved molecules of high dielectric constant (highly polar molecules where $\varepsilon_2 > \varepsilon$) but repelled from molecules of low dielectric constant (nonpolar molecules where $\varepsilon_2 < \varepsilon$).

(iii)   The interaction between any two identical uncharged molecules ($\varepsilon_1 = \varepsilon_2$) is always attractive regardless of the nature of the solvent medium. Interestingly, two microscopic air bubbles also attract each other in a medium.

This approach will generally predict the right qualitative trends, but it is quantitatively somewhat model-dependent in that it treats molecules as if they were a uniform medium having bulk dielectric properties. It is essentially a *continuum* treatment where the molecular properties only appear in determining the radius or volume of the molecules. This may be valid for larger molecules, macromolecules, and small particles in solution but may fail for small molecules. Let us therefore end this chapter by looking into the reasonableness of treating a molecule as if it were a dielectric sphere. Now from Eq. (5.26) we see that for isolated molecules in the gas phase ($\varepsilon = 1$), their *total* polarizability should be given by

$$\frac{\alpha}{(4\pi\varepsilon_0)} = \left(\frac{\varepsilon - 1}{\varepsilon + 2}\right)\frac{3v}{4\pi}, \qquad (5.28)$$

where ε is now the static dielectric constant of the molecules, assumed to be the same as that of the condensed state (e.g., the liquid state). Equation (5.28) is known as the Clausius–Mossotti equation. If we are interested in the *electronic* polarizability α_0, then the value of ε in Eq. (5.28) is that measured in the visible range of frequencies and equals n^2, where n is the refractive index of the medium. Thus, for the *electronic* polarizability, we may write

$$\frac{\alpha_0}{(4\pi\varepsilon_0)} = \left(\frac{n^2 - 1}{n^2 + 2}\right)\frac{3v}{4\pi}, \tag{5.29}$$

which is known as the Lorenz–Lorentz equation.

Table VII shows the experimental values for the electronic polarizabilities α_0 and dipole moments u of isolated molecules in the gas phase, from which their values of α may be obtained from Eq. (5.7):

$$\alpha = \alpha_0 + u^2/3kT.$$

Table VII also shows the polarizabilities α_0 and α as calculated from Eqs. (5.29) and (5.28) in terms of the purely bulk properties of the liquids: the refractive index n, the dielectric constant ε, and the molecular weight M, and mass density ρ from which the volume occupied per molecule is given by

$$v = M/\rho N_0, \tag{5.30}$$

where N_0 is Avogadro's constant.

It is evident that the electronic polarizabilities α_0 of all the molecules listed in Table VII are excellently described by Eq. (5.29), whereas the total polarizabilities α of molecules are well described by Eq. (5.28) only for weakly polar molecules such as $CHCl_3$ and C_6H_5OH. For small highly polar molecules such as H_2O and $(CH_3)_2CO$, the agreement is not good but improves for progressively larger molecules (cf. $CH_3OH \rightarrow C_2H_5OH \rightarrow C_6H_{13}OH$ and C_6H_5OH) as expected. It is not possible to ascertain whether the lack of agreement for highly polar molecules is due to the inapplicability of Eq. (5.28) or to a changed dipole moment of the molecules in the liquid state. However, we may safely conclude that Eqs. (5.28)–(5.29) serve as quantitatively reliable equations for determining the polarizabilities of all but very small highly polar molecules, and we may expect that Eq. (5.26) for the excess polarizability and Eq. (5.27) for the free energy should be likewise applicable.

But we are still not in a position to estimate the strength of the total interaction between neutral or polar molecules. There is one final contribution to the total force between *all* molecules and particles that must be considered before we can do that. This is the van der Waals–dispersion force.

TABLE VII
Molecular polarizabilities as determined from molecular or bulk properties[a,b]

Molecule		Polarizabilities from Gas (Molecular Properties)			Polarizabilities from Condensed Phase (Continuum) Properties					
		$\dfrac{\alpha_0}{(4\pi\varepsilon_0)}$ $(10^{-30}\,\text{m}^3)$	u_{gas} (D)	$\dfrac{\alpha}{(4\pi\varepsilon_0)}$ from $\alpha=\alpha_0+\dfrac{u^2}{3kT}$ $(10^{-30}\,\text{m}^3)$	M $(10^{-3}\,\text{kg mol}^{-1})$	ρ $(10^{-3}\,\text{kg m}^{-3})$	n	ε	$\dfrac{\alpha_0}{(4\pi\varepsilon_0)}$ from $\left(\dfrac{n^2-1}{n^2+2}\right)\dfrac{3M}{4\pi N_0\rho}$ $(10^{-30}\,\text{m}^3)$	$\dfrac{\alpha}{(4\pi\varepsilon_0)}$ from $\left(\dfrac{\varepsilon-1}{\varepsilon+2}\right)\dfrac{3M}{4\pi N_0\rho}$ $(10^{-30}\,\text{m}^3)$
CCl_4	carbon tetrachloride	10.5	0	10.5	153.8	1.59	1.460	2.24	10.5	11.2
C_6H_6	benzene	10.3	0	10.3	78.1	0.88	1.501	2.28	10.4	10.5
$CHCl_3$	chloroform	8.2	1.06	17.5	119.4	1.48	1.446	4.81	8.5	17.9
H_2O	water	1.5	1.85	29.7	18.0	1.00	1.333	80	1.47	6.9
$(CH_3)_2CO$	acetone	6.4	2.85	73.4	58.1	0.79	1.359	21	6.4	25.3
CH_3OH	methanol	3.2	1.69	26.8	32.0	0.79	1.329	33	3.3	14.7
C_2H_5OH	ethanol	5.2	1.69	28.8	46.1	0.79	1.361	26	5.1	20.7
$n\text{-}C_6H_{13}OH$	hexanol	12.5	1.69	36.1	102.2	0.81	1.418	13	12.6	40.0
C_6H_5OH	phenol	11.2	1.45	26.4	94.1	1.07	1.551	10 (60°)	11.1	26.1

[a] All values at 20°C unless stated otherwise.
[b] α_0, electronic polarizability; α, total polarizability.

Chapter 6

van der Waals Forces

6.1. Origin of the van der Waals–dispersion force between neutral molecules: the London equation

The various types of physical forces so far described are fairly easy to understand since they arise from straightforward electrostatic interactions involving charged or dipolar molecules. But there is a further type of force, which—like the gravitational force—acts between *all* atoms and molecules, even totally neutral ones such as helium, carbon dioxide, and hydrocarbons. These forces have been variously known as dispersion forces, London forces, charge-fluctuation forces, electrodynamic forces, and induced-dipole–induced-dipole forces. We shall refer to them as *dispersion forces* since it is by this name that they are most widely known. The origin of this name has to do with their relation to the dispersion of light in the visible and UV regions of the spectrum, as we shall see. The literature on this subject is quite voluminous, and the reader is referred to books and reviews by London (1937), Hirschfelder *et al.* (1954), Moelwyn-Hughes (1961), Margenau and Kestner (1971), Israelachvili and Tabor (1973), Israelachvili (1974), and Mahanty and Ninham (1976).

Dispersion forces make up the third and perhaps most important contribution to the total van der Waals force between atoms and molecules, and because they are always present (in contrast to the other types of forces that may or may not be present depending on the properties of the molecules) they play a role in a host of important phenomena such as adhesion, surface tension, physical adsorption, the properties of gases and liquids, the strengths of solids, the flocculation or aggregation of particles in aqueous solutions, and the structures of condensed macromolecules such as proteins and polymers.

65

Their main features may be summarized as follows:

(1)   They are long-range forces and, depending on the situation, can be effective from large distances (greater than 10 nm) down to interatomic spacings (about 0.2 nm).

(2)   These forces may be repulsive or attractive, and in general the dispersion force between two molecules or large particles does not follow a simple power law.

(3)   Dispersion forces not only bring molecules together but also tend to mutually align or orient them, though this orienting effect is usually weak.

(4)   The dispersion interaction of two bodies is affected by the presence of other bodies nearby. This is known as the *nonadditivity* of an interaction.

Dispersion forces are quantum mechanical in origin and amenable to a host of theoretical treatments of varying complexity, the most rigorous of which would take us into the world of quantum electrodynamics. Their origin may be understood intuitively as follows: For a nonpolar atom such as helium, the time average of its dipole moment is zero, yet at any instant there exists a finite dipole moment given by the instantaneous positions of the electrons about the nuclear protons. This instantaneous dipole generates an electric field that polarizes any nearby neutral atom, inducing a dipole moment in it. The resulting interaction between the two dipoles gives rise to an instantaneous attractive force between the two atoms, and the time average of this force is finite. Now for a simple semiquantitative understanding of how these forces arise, let us consider the following model based on the interaction between two *Bohr atoms*.

In the Bohr atom an electron is pictured as orbiting around a proton. The smallest distance between the electron and proton is known as the *first Bohr radius* a_0 and is the radius at which the Coulomb energy $e^2/4\pi\varepsilon_0 a_0$ is equal to $2h\nu$, that is,

$$a_0 = e^2/2(4\pi\varepsilon_0)h\nu = 0.053 \text{ nm}, \qquad (6.1)$$

where h is the Planck constant and ν the orbiting frequency of the electron. For a Bohr atom, $\nu = 3.3 \times 10^{15} \text{ s}^{-1}$, so that $h\nu = 2.2 \times 10^{-18} \text{ J}$. This is the energy of an electron in the first Bohr radius and is equal to the energy needed to ionize the atom—the *first ionization potential I*.

The Bohr atom has no permanent dipole moment. However, at any instant there exists an instantaneous dipole of moment

$$u = a_0 e$$

whose field will polarize a nearby neutral atom giving rise to an attractive interaction that is entirely analogous to the dipole-induced dipole interaction discussed in Chapter 5. The energy of this interaction in vacuum will therefore be given by Eq. (5.22) as

$$w(r) = -u^2\alpha_0/(4\pi\varepsilon_0)^2 r^6 = -(a_0 e)^2 \alpha_0/(4\pi\varepsilon_0)^2 r^6,$$

where α_0 is the electronic polarizability of the (second) Bohr atom, which from Eq. (5.4) is approximately

$$\alpha_0 \approx 4\pi\varepsilon_0 a_0^3.$$

Using this expression for α_0 and Eq. (6.1) for a_0, we immediately find that the above interaction energy can be written approximately as

$$w(r) \approx -\alpha_0^2 h\nu/(4\pi\varepsilon_0)^2 r^6. \tag{6.2}$$

Except for a numerical factor Eq. (6.2) is the same as that derived by London in 1930 using quantum mechanical perturbation theory. London's famous expression for the dispersion interaction energy between two identical atoms or molecules is (London, 1937)

$$w(r) = -\tfrac{3}{4}\alpha_0^2 h\nu/(4\pi\varepsilon_0)^2 r^6 = -\tfrac{3}{4}\alpha_0^2 I/(4\pi\varepsilon_0)^2 r^6, \tag{6.3}$$

and for two dissimilar atoms,

$$w(r) = -\frac{3}{2}\frac{\alpha_{01}\alpha_{02}}{(4\pi\varepsilon_0)^2 r^6}\frac{h\nu_1\nu_2}{(\nu_1 + \nu_2)} = -\frac{3}{2}\frac{\alpha_{01}\alpha_{02}}{(4\pi\varepsilon_0)^2 r^6}\frac{I_1 I_2}{(I_1 + I_2)}. \tag{6.4}$$

London's equation has since been superceded by more exact though more complicated expressions (see Section 6.5), but it can be relied upon to give fairly accurate values though these are usually lower than more rigorously determined ones.

From the above simple model we see that while dispersion forces are quantum mechanical (in determining the instantaneous, but fluctuating, dipole moments of neutral atoms), the ensuing interaction is still essentially electrostatic—a sort of quantum mechanical polarization force. And we may further note that the $1/r^6$ distance dependence is the same as that for the other two polarization interactions (the Keesom and Debye interactions) that contribute to the net van der Waals force, discussed in Section 5.6. But before we consider these three interactions collectively let us first investigate the nature of dispersion forces.

6.2. Strength of dispersion forces: van der Waals solids and liquids

As an estimate of the strength of the dispersion energy, we may consider two atoms or small molecules with $\alpha_0/4\pi\varepsilon_0 \approx 1.5 \times 10^{-30}$ m^3 and $I = h\nu \approx 2 \times 10^{-18}$ J (a typical ionization potential). From Eq. (6.3) we find that for two atoms in contact at $r = \sigma \approx 0.3$ nm, $w(\sigma) = -4.6 \times 10^{-21}$ J $\approx 1\,kT$. This is a very respectable energy, considering that the interaction appears at first sight to spring up from nowhere. But when we recall that the inducing (instantaneous) dipole moment of even a small hydrogen (Bohr) atom is of order $a_0 e \approx 2.4$ D, we can appreciate why the dispersion interaction is by no means negligible. Thus, while very small nonpolar atoms and molecules such as argon and methane are gaseous at room temperature and pressure, larger molecules such as hexane and higher molecular weight hydrocarbons are liquids or solids, held together solely by dispersion forces. These solids are referred to as *van der Waals* solids, and they are characterized by having weak undirected "bonds," and therefore low melting points and low latent heats of melting.

For spherically symmetrical inert molecules such as neon, argon, and methane, the van der Waals solids they form at low temperatures are close packed structures with 12 nearest neighbours per atom. Their lattice energy (12 shared "bonds" or six full "bonds" per atom) is therefore approximately $6w(\sigma)$ per molecule, though if the attractions of more distant neighbours are also included the factor of 6 rises to 7.22. The expected *molar lattice energy* or *cohesive energy* of a van der Waals solid is therefore

$$U \approx 7.22\, N_0 \left[\frac{3\alpha_0^2 h\nu}{4(4\pi\varepsilon_0)^2 \sigma^6} \right], \qquad (6.5)$$

where σ is now the equilibrium interatomic distance in the solid. Thus, for argon, since $\alpha_0/(4\pi\varepsilon_0) = 1.63 \times 10^{-30}$ m^3, $I = h\nu = 2.52 \times 10^{-18}$ J, and $\sigma = 0.376$ nm, we obtain $U \approx 7.7$ kJ mol^{-1}, which may be compared with the latent heat of melting plus vapourization for argon of $L_m + L_v = 7.7$ kJ mol^{-1}, which is approximately equal to the latent heat of sublimation, or the cohesive energy. Table VIII shows the calculated and experimental values for the cohesive energies U of a number of van der Waals solids composed of small spherical atoms or molecules. The good agreement obtained is to some extent fortuitous since the computed values neglect

(i)   the very short-range stabilizing repulsive forces (Chapter 7) that *lower* the final binding energy at contact (by up to 50%), and

(ii)   other attractive forces, such as arise from fluctuating quadrupole interactions and *many-body effects* (Section 6.8), which *raise* the final binding (by up to 50%).

These two opposing effects partially cancel each other so that the final results look more impressive than they deserve.

For larger spherical molecules, such as CCl_4, the simple London equation can no longer be used. This is because the dispersion force no longer acts

TABLE VIII

Strength of dispersion interaction between spherical nonpolar molecules

Interacting Molecules	Molecular Diameter σ (nm) (From Fig. 14)	Polarizability $\alpha_0/4\pi\varepsilon_0$ (10^{-30} m³)	Ionization Potential $I = h\nu_I$ (eV)	$3\alpha_0^2 h\nu_I/4(4\pi\varepsilon_0)^2$ (10^{-79} J m⁶)		Molar Cohesive Energy U (kJ mol⁻¹)		Boiling Point T_B (K)	
				Theoretical	From Gas Law[a] Eq. (6.14)	Theoretical Eq. (6.5)	Measured (Approx.) $L_m + L_v$	Measured	$3\alpha_0^2 h\nu_I/4(4\pi\varepsilon_0)^2\sigma^6$ (1.5k)
Ne–Ne	0.308	0.39	21.6	3.9	3.8	2.0	2.1	27	22
Ar–Ar	0.376	1.63	15.8	50	45	7.7	7.7	87	85
CH₄–CH₄	0.400	2.60	12.6	102	101	10.9	9.8	112	121
Xe–Xe	0.432	4.01	12.1	233	225	15.6	14.9	165	173
CCl₄–CCl₄	0.550	10.5	11.5	1520	2960	23.9	32.6	350	265

[a] van der Waals constants a and b taken from the "Handbook of Chemistry and Physics," CRC Press, 56th ed.

between the centres of the molecules, but between the centres of electronic polarization within each molecule, which, as we saw in Chapter 5, are located at the covalent bonds (clearly for very large spheres the centre-to-centre distance ceases to have any direct significance as regards the strength of the cohesion energy). Thus, for CCl_4, the calculated value for the cohesion energy is too small (Table VIII) because the distance between the polarizable electrons between two molecules is now significantly less than the equilibrium intermolecular separation of 0.55 nm. In Chapter 11 we shall investigate the interactions between spheres and particles whose radii are much larger than interatomic spacings.

Likewise, the simple London equation or Eq. (6.5) cannot be applied to asymmetric (nonspherical) molecules such as alkanes, polymers, and cyclic or planar molecules. To compute the binding energies within or between such complex molecules the molecular packing in the solid or liquid must be known (which, of course, transforms any theoretical endeavour from the predictive to the confirmatory), and the different contributions arising from different parts of the molecules must be taken into account. Under such conditions the exact dispersion interaction is difficult to compute, but some simplifying assumptions can often be made to arrive at reasonable working models. Let us consider one such model for normal alkanes, of general formula CH_3—$(CH_2)_n$—CH_3, where each molecule may be considered as a cylinder of diameter $\sigma = 0.40$ nm composed of CH_2 groups spaced linearly at intervals of $l = 0.127$ nm, corresponding to the normal CH_2—CH_2 distance along an alkane chain. If we now consider one such molecule surrounded by six close-packed neighbouring cylinders, we may then sum the dispersion energy of any one CH_2 group in the central molecule with all the CH_2 groups in the six surrounding molecules (similar to the lattice sum carried out in Section 3.4 for the ionic lattice energy). Thus, there will be 6 CH_2 groups at $r = \sigma$, 12 at $r = [\sigma^2 + l^2]^{1/2}$, 12 at $r = [\sigma^2 + (2l)^2]^{1/2}$, and so on. The molar cohesive energy per CH_2 group will therefore be given by the following rapidly converging series:

$$U = \frac{3\alpha_0^2 h\nu}{4(4\pi\varepsilon_0)^2} \left[\frac{6}{\sigma^6} + \frac{12}{[\sigma^2 + l^2]^3} + \frac{12}{[\sigma^2 + (2l)^2]^3} + \cdots \right] \frac{N_0}{2}. \qquad (6.6)$$

Now for CH_2 groups, $\alpha_0/4\pi\varepsilon_0 = 1.84 \times 10^{-30}$ m^3, and $h\nu \approx 10.4$ eV, so that we obtain

$$U \approx 6.9 \quad \text{kJ mol}^{-1} \text{ per } CH_2 \text{ group.} \qquad (6.7)$$

Table IX shows the experimental values for the latent heats of melting plus vapourization of alkanes together with the computed values based on Eq. (6.7). Experimentally it is found that for straight chained alcohols, carboxylic acids, amides, esters, etc., the cohesive energy increases by between 6

TABLE IX
Strength of dispersion interaction between nonspherical
alkane molecules

Molecule	Number of CH$_2$ Groups n	Molar Cohesive Energy (kJ mol^{-1})	
		Theoretical Eq. (6.7)	Measured $L_m + L_v$
CH$_4$	0	9.8 (measured)	9.8
C$_6$H$_{14}$	5	44.3	45.0
C$_{12}$H$_{26}$	11	85.7	86.1
C$_{18}$H$_{38}$	17	127.1	125.9

and 7.5 kJ mol^{-1} per added CH$_2$ group. These good agreements between theory and experiment show two important aspects of dispersion forces: their nondirectionality and nearadditivity, though for interactions across a medium the dispersion force can be very nonadditive (see Sections 6.6 and 6.8).

6.3. van der Waals equation of state

Let us now see how the dispersion force potential can be related to the constants a and b in the van der Waals equation of state

$$(P + a/V^2)(V - b) = RT. \tag{6.8}$$

We shall again consider the molecules to be *hard spheres* of diameters σ, whose pair interaction energy is given by

$$w(r) = -\frac{3\alpha_0^2 h\nu}{4(4\pi\varepsilon_0)^2 r^6} = -\frac{C}{r^6}, \qquad r > \sigma,$$

$$= \infty, \qquad r < \sigma. \tag{6.9}$$

In Section 2.3 we found that the constant a is given by

$$a = 2\pi C/(n-3)\sigma^{n-3} = 2\pi C/3\sigma^3 \qquad \text{for} \quad n = 6.$$

This value, however, was derived for the case where the van der Waals equation was expressed in terms of molecular parameters [Eq. (2.10)]: $(P + a/v^2)(v - b) = kT$. Since $v = V/N_0$ we see that when V is the molar volume of the gas then

$$a = 2\pi N_0^2 C/3\sigma^3. \tag{6.10}$$

The London dispersion force coefficient C is therefore related to the van der Waals constant a by

$$C = 3\sigma^3 a/2\pi N_0^2. \tag{6.11}$$

The constant b is obtained from the volume unavailable for the molecules to move in, the "excluded volume" per mole. Since one molecule cannot approach closer than σ to another, the excluded volume for the pair is $\frac{4}{3}\pi\sigma^3$, or $\frac{2}{3}\pi\sigma^3$ per molecule as previously derived in Section 2.3. Thus,

$$b = \tfrac{2}{3}\pi N_0 \sigma^3 \tag{6.12}$$

per mole, which is four times the molar volume of the molecules. We therefore have

$$\sigma^3 = 3b/2\pi N_0. \tag{6.13}$$

Combining the above with Eq. (6.11) we may finally write for C,

$$C = 9ab/4\pi^2 N_0^3 = 1.05 \times 10^{-76} ab \quad \text{J m}^6, \tag{6.14}$$

where a is in dm^6 atm mol^{-2} and b in dm^3 mol^{-1}. For example, for methane, CH_4, the experimental values are $a = 2.25$ dm^6 atm mol^{-2}, and $b = 0.0428$ dm^3 mol^{-1}, giving $C = 101 \times 10^{-79}$ J m^6, which is in remarkably good agreement with the value for C of $3\alpha_0^2 h\nu/4(4\pi\varepsilon_0)^2 = 102 \times 10^{-79}$ J m^6 calculated on the basis of the London equation. Table VIII also shows how good this agreement is for the other molecules listed, except for the largest molecule CCl_4, which, as already discussed, must have a stronger interaction than could be accounted for by applying the London equation between molecular centres.

For molecules constrained to interact on a surface, as occurs on adsorption and in surface monolayers, there is an analogous equation to the van der Waals equation of state. This two-dimensional analogue may be written as

$$(\Pi + a/A^2)(A - b) = RT, \tag{6.15}$$

where Π is the external surface pressure (in N m^{-1}), A the area, and a and b are constants as before. It is easy to verify that for an intermolecular pair potential of the form $w(r) = -C/r^6$, the constants a and b now become

$$a = \pi C N_0^2/4\sigma^4, \qquad b = \tfrac{1}{2}\pi N_0 \sigma^2. \tag{6.16}$$

Finally, we may note that for purely repulsive forces, the sign of a changes, and the resulting equations then predict a monotonic decrease of V with P, or

A with Π, with no gas to liquid transition (i.e., no boiling temperature). But at high—near close-packed—densities there is always a melting transition from a disordered (liquid) state to an ordered quasi-crystalline (solid) state (Alder *et al.*, 1968). The different roles played by attractive and repulsive forces in determining the melting and boiling points of substances is further discussed in Section 7.4.

6.4. van der Waals forces between polar molecules

Three distinct types of force contribute to the total long-range interaction between polar molecules, collectively known as the van der Waals force: these are the *induction* force, the *orientation* force, and the *dispersion* force, each of which has an interaction free energy that varies with the inverse sixth power of the distance. Thus, for two dissimilar polar molecules, we have

$$w_{\mathrm{VDW}}(r) = -C_{\mathrm{VDW}}/r^6 = -[C_{\mathrm{ind}} + C_{\mathrm{orient}} + C_{\mathrm{disp}}]/r^6$$

$$= -\left[(u_1^2 \alpha_{02} + u_2^2 \alpha_{01}) + \frac{u_1^2 u_2^2}{3kT} + \frac{3\alpha_{01}\alpha_{02}h\nu_1\nu_2}{2(\nu_1 + \nu_2)} \right] \Big/ (4\pi\varepsilon_0)^2 r^6. \quad (6.17)$$

Table X shows how these three forces contribute to the net van der Waals energies of some molecules. The table reveals some interesting and important properties of van der Waals interactions:

(i) *Dominance of dispersion forces.* Dispersion forces generally exceed the dipole-dependent induction and orientation forces except for small highly polar molecules, such as water. The relative unimportance of dipolar forces is clearly seen in the hydrogen halides: as we go from HCl to HI the strength of the van der Waals interaction increases even as the dipole moments diminish. Note, too, that in the interaction between two dissimilar molecules of which one is nonpolar, the van der Waals energy is almost completely dominated by the dispersion contribution.

(ii) *Comparisons with experimental data.* The agreements between the computed (theoretical) values for C_{VDW} and those obtained from the gas law coefficients *a* and *b* are surprisingly good, even for NH_3 and H_2O. It is also possible to estimate the molar cohesive energies of polar molecules (London, 1937), though the agreement is not as good as for the nonpolar spherical molecules listed in Table VIII partly because the effective diameters σ of polar (and therefore asymmetrical) molecules are not well defined. For example, for CH_3Cl of molecular diameter $\sigma \approx 0.43$ nm, using Eq. (6.5) we obtain $U \approx 29$ kJ mol^{-1} compared to the experimental value for $L_m + L_v$ of ~26 kJ mol^{-1}, while for water ($\sigma = 0.28$ nm) with only four

TABLE X
Induction, orientation, and dispersion free energy contributions to the total van der Waals energy in vacuum for various pairs of molecules at 293 K

Interacting Molecules	Electronic Polarizability $\frac{\alpha_0}{4\pi\varepsilon_0}$ (10^{-30} m³)	Permanent Dipole Moment u (D)[a]	Ionization Potential $I = h\nu_1$ (eV)[b]	van der Waals energy coefficients C (10^{-79} J m⁶) C_{ind} $\dfrac{2u^2\alpha_0}{(4\pi\varepsilon_0)^2}$	C_{orient} $\dfrac{u^4/3kT}{(4\pi\varepsilon_0)^2}$	C_{disp} $\dfrac{3\alpha_0^2 h\nu_1}{4(4\pi\varepsilon_0)^2}$	Total VdW Energy C_{vdw} Theoretical Eq. (6.17)	From Gas Law Eq. (6.14)	Dispersion Energy Contribution To Total (Theoretical) (%)
Ne–Ne	0.39	0	21.6	0	0	4	4	4	100
CH$_4$–CH$_4$	2.60	0	12.6	0	0	102	102	101	100
HCl–HCl	2.63	1.08	12.7	6	11	106	123	157	86
HBr–HBr	3.61	0.78	11.6	4	3	182	189	207	96
HI–HI	5.44	0.38	10.4	2	0.2	370	372	350	99
CH$_3$Cl–CH$_3$Cl	4.56	1.87	11.3	32	101	282	415	509	68
NH$_3$–NH$_3$	2.26	1.47	10.2	10	38	63	111	162	57
H$_2$O–H$_2$O	1.48	1.85	12.6	10	96	33	139	175	24

Dissimilar Molecules

Interacting Molecules	C_{ind} $\dfrac{u_1^2\alpha_{02}+u_2^2\alpha_{01}}{(4\pi\varepsilon_0)^2}$	C_{orient} $\dfrac{u_1^2u_2^2/3kT}{(4\pi\varepsilon_0)^2}$	C_{disp} $\dfrac{3\alpha_{01}\alpha_{02}h\nu_1\nu_2}{2(4\pi\varepsilon_0)^2(\nu_1+\nu_2)}$	Theoretical Eq. (6.17)	From Gas Law Eq. (6.14)	Dispersion Energy Contribution To Total (%)
Ne–CH$_4$	0	0	19	19	—	100
HCl–HI	7	1	197	205	—	96
H$_2$O–Ne	1	0	11	12	—	92
H$_2$O–CH$_4$	9	0	58	67	—	87

[a] 1 D = 3.336 × 10^{-30} C m.
[b] 1 eV = 1.602 × 10^{-19} J.

nearest neighbours per molecule (see Chapter 8) we find $U \approx 70$ kJ mol^{-1} compared to the measured value of ~ 50 kJ mol^{-1}.

(iii) *Interactions of dissimilar molecules*. The van der Waals interaction energy between two dissimilar molecules A and B is usually intermediate between the values for A–A and B–B. In fact, the coefficient C_{VDW} for A–B is often close to the geometric mean of A–A and B–B. Thus, for Ne–CH$_4$, the geometric mean (see Table X) is $\sqrt{4 \times 102} = 20$, which may be compared with the computed value of 19, while for HCl–HI we obtain $\sqrt{123 \times 372} = 214$ compared to the computed value of 205. This procedure affords a convenient way of estimating the van der Waals interactions between unlike molecules when direct experimental data is not available. However, for interactions involving water, it breaks down, and for H$_2$O–CH$_4$, the net interaction is actually less than that of H$_2$O–H$_2$O or CH$_4$–CH$_4$. Thus, water and methane are attracted to themselves more strongly than they are attracted to each other. It is partly for this reason that nonpolar molecules are not miscible (do not mix) with water but separate out into different phases in water. Such compounds (hydrocarbons, fluorocompounds, oils, fats) are known as *hydrophobic* (from the Greek "water fearing"), and their low water solubility and their propensity to separate into clusters or phases in water is known as the *hydrophobic effect*. Table X, however, only reveals part of the story: in liquid water the temperature-dependent orientation (dipole–dipole) interaction and the induction (dipole–neutral molecule) interaction are greatly modified, as will be discussed later.

6.5. General theory of van der Waals forces between molecules

The London theory of dispersion forces has two serious shortcomings. It assumes that atoms and molecules have only a single ionization potential (one absorption frequency), and it cannot handle the interactions of molecules in a solvent. In 1963 McLachlan presented a generalized theory of van der Waals forces, which included in one equation the induction, orientation, and dispersion force, and which may also be readily applied to interactions in a solvent medium. McLachlan's expression for the van der Waals free energy of two molecules or small particles 1 and 2 in a medium 3 is given by the series (McLachlan, 1963a,b)

$$w(r) = -\frac{6kT}{(4\pi\varepsilon_0)^2 r^6} \sum_{n=0,1,2,\ldots}^{\infty}{}' \frac{\alpha_1(iv_n)\alpha_2(iv_n)}{\varepsilon_3^2(iv_n)}, \qquad (6.18)$$

where $\alpha_1(iv_n)$ and $\alpha_2(iv_n)$ are the polarizabilities of molecules 1 and 2 and $\varepsilon_3(iv_n)$

the dielectric permittivity of the medium 3 at *imaginary* frequencies iv_n, where*

$$v_n = (2\pi kT/h) \approx 4 \times 10^{13}n \quad s^{-1} \quad \text{at 300 K} \tag{6.19}$$

and where the prime over the summation (Σ') indicates that the zero frequency $n = 0$ term is multiplied by $\frac{1}{2}$. McLachlan's equation looks complicated, but it is actually quite straightforward to compute once we realize that $\alpha(iv_n)$ and $\varepsilon(iv_n)$ are real quantities, easily related to measurable properties, as we shall now see. Thus, for a molecule with one absorption frequency (the ionization frequency) v_I, its electronic polarizability at *real* frequencies v is given by the simple harmonic oscillator model (Von Hippel, 1958)

$$\alpha(v) = \alpha_0/[1 - (v/v_I)^2]. \tag{6.20}$$

Note that since most absorption frequencies are in the ultraviolet region (typically, $v_1 \approx 3 \times 10^{15} \ s^{-1}$) the value of $\alpha(v_{vis})$ in the visible range of frequencies $v_{vis} \approx 5 \times 10^{14} \ s^{-1}$ is essentially the same as α_0 because $(v_{vis}/v_I)^2 \ll 1$.

If the molecule also possesses a permanent dipole moment u there is an additional *orientational* or dipole relaxation polarizability contribution (Von Hippel, 1958) given by

$$\alpha_{orient}(v) = u^2/3kT(1 - iv/v_{rot}), \tag{6.21}$$

where v_{rot} is some average rotational relaxation frequency for the molecule, usually in the far infrared or microwave region of frequencies (typically, $v_{rot} \approx 10^{11} \ s^{-1}$).

The total polarizability of a molecule in free space as a function of iv_n (replacing v by iv_n) may therefore be written as

$$\alpha(iv_n) = \frac{u^2}{3kT(1 + v_n/v_{rot})} + \frac{\alpha_0}{1 + (v_n/v_I)^2}. \tag{6.22}$$

At zero frequency $(v_n = 0)$ this reduces to the Debye–Langevin equation, Eq. (5.7):

$$\alpha(0) = u^2/3kT + \alpha_0. \tag{6.23}$$

The variation of $\alpha(iv)$ with frequency v for a simple polar molecule is shown schematically in Fig. 12.

We may now return to McLachlan's expression, Eq. (6.18). The first term in the series is that for $n = 0$ (i.e., $v_n = 0$) so that from the above we immediately

---

* In the literature, frequencies are often denoted by ω where $\omega = 2\pi v$, and Planck's constant by h where $\hbar = h/2\pi$.

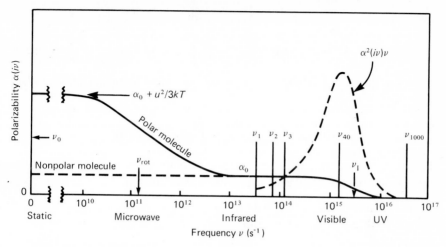

Fig. 12 The polarizability $\alpha(iv)$ as a function of frequency v of a simple polar molecule is given by Eq. (6.22):

$$\alpha(iv) = \frac{u^2}{3kT(1 + v/v_{\text{rot}})} + \frac{\alpha_0}{1 + (v/v_1)^2}.$$

The frequencies $v_n = 4 \times 10^{13} n$ s^{-1} at which $\alpha(iv)$ contributes to the van der Waals energy (at 300 K) are shown as a series of vertical lines. The curve of $\alpha^2(iv)v$ shows how different regions of the spectrum contribute to the dispersion energy.

obtain the "zero frequency contribution" to $w(r)$,

$$w(r)_{v=0} = -\frac{3kT}{(4\pi\varepsilon_0)^2 r^6}\alpha_1(0)\alpha_2(0)$$

$$= -\frac{3kT}{(4\pi\varepsilon_0)^2 r^6}\left[\frac{u_1^2}{3kT} + \alpha_{01}\right]\left[\frac{u_2^2}{3kT} + \alpha_{02}\right], \qquad (6.24)$$

which is identical to Eq. (5.25) derived in Section 5.6 and includes both the orientation (Keesom) and induction (Debye) interaction energies discussed earlier.

Let us now turn to the nonzero frequency terms ($n = 1, 2, 3, \ldots$) of Eq. (6.18). The summation entails calculating the values of $\alpha(iv_n)$ at the discrete frequencies given by Eq. (6.19). Now at normal temperatures the first frequency $v_{n=1} \approx 4 \times 10^{13}$ s^{-1} is already much greater than v_{rot}, so that $\alpha(iv_n)$ is effectively determined solely by the electronic polarizability contribution in Eq. (6.22). Further, since $v_{n=1}$ is still much smaller than a typical absorption frequency $v_1 \approx 3 \times 10^{15}$ s^{-1}, it is clear that the frequencies v_n are very close together in the UV region. We may therefore replace the sum of discrete frequencies Σ by an integration over n (i.e., $dn = (h/2\pi kT)\,dv$) so that

$$kT \sum_{n=1,2,\ldots}^{\infty} \rightarrow \frac{h}{2\pi} \int_{v_1}^{\infty} dv. \qquad (6.25)$$

Applying this to Eq. (6.18) we obtain for the "finite frequency" free energy contribution to $w(r)$

$$w(r)_{v>0} = -\frac{3h}{(4\pi\varepsilon_0)^2\pi r^6} \int_{v_1}^{\infty} \alpha_1(iv)\alpha_2(iv)\,dv. \qquad (6.26)$$

Finally, by substituting the electronic polarizability as expressed by Eq. (6.22) into the above and integrating using the definite integral

$$\int_0^{\infty} \frac{dx}{(a^2 + x^2)(b^2 + x^2)} = \frac{\pi}{2ab(a + b)}, \qquad (6.27)$$

we obtain

$$w(r)_{v>0} \approx -\frac{3\alpha_{01}\alpha_{02}}{2(4\pi\varepsilon_0)^2 r^6} \frac{h v_{11} v_{12}}{(v_{11} + v_{12})}, \qquad (6.28)$$

which is the London equation. The complete McLachlan formula is particularly suitable for computing the dispersion force between molecules that have a number of different absorption frequencies or ionization potentials, for which the simple London expression breaks down.

6.6. van der Waals forces in a medium

The theory of McLachlan is also naturally applicable to the interactions of molecules or small particles in a medium (McLachlan, 1965). In this case the polarizabilities $\alpha(iv)$ in Eq. (6.18) are now the *excess* or *effective polarizabilities* of the molecules as discussed in Section 5.7, and for a small spherical molecule 1 of radius a_1 in a medium 3 (Fig. 13a), the excess polarizability is given approximately by

$$\alpha_1(v) = 4\pi\varepsilon_0\varepsilon_3\left(\frac{\varepsilon_1(v) - \varepsilon_3(v)}{\varepsilon_1(v) + 2\varepsilon_3(v)}\right)a_1^3. \qquad (6.29)$$

The zero-frequency contribution to $w(r)$ in Eq. (6.18) is therefore

$$w(r)_{v=0} = -\frac{3kTa_1^3 a_2^3}{r^6}\left(\frac{\varepsilon_1(0) - \varepsilon_3(0)}{\varepsilon_1(0) + 2\varepsilon_3(0)}\right)\left(\frac{\varepsilon_2(0) - \varepsilon_3(0)}{\varepsilon_2(0) + 2\varepsilon_3(0)}\right), \qquad (6.30)$$

where $\varepsilon_1(0)$, $\varepsilon_2(0)$, and $\varepsilon_3(0)$ are the (static) dielectric constants of the three media. Equation (6.30) is the same as that previously derived and discussed in

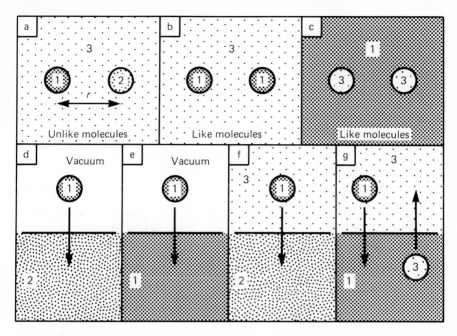

Fig. 13 (a)–(c):   Molecules of radii a (diameter $\sigma = 2a$) interacting in a solvent medium.
(d)–(g):   Transfer of molecules from one medium to another.

Section 5.7 in connection with the orientation and induction forces in a medium.

We shall return to it again after first considering the (finite frequency) dispersion contribution. Substituting Eq. (6.29) into Eq. (6.18), and replacing the summation by the integral of Eq. (6.25), this may be written as

$$w(r)_{v>0} = -\frac{3ha_1^3 a_2^3}{\pi r^6} \int_0^\infty \left(\frac{\varepsilon_1(iv) - \varepsilon_3(iv)}{\varepsilon_1(iv) + 2\varepsilon_3(iv)}\right)\left(\frac{\varepsilon_2(iv) - \varepsilon_3(iv)}{\varepsilon_2(iv) + 2\varepsilon_3(iv)}\right)dv. \quad (6.31)$$

Unfortunately, complete data on the absorption spectra of most materials is not available, and it is necessary to adopt a model that suitably represents the dielectric behaviour of the media. In the previous section we saw how it was possible to derive London's result for the dispersion force by treating molecules as simple harmonic oscillators. Similarly, for a dielectric medium that shows one strong absorption peak at a frequency v_e (which is usually different from that of isolated molecules in the gas v_1) we may express its dielectric permittivity at frequencies $v > v_1$ as (Von Hippel, 1958)

$$\varepsilon(v) = 1 + (n^2 - 1)/[1 - (v/v_e)^2] \quad (6.32)$$

or

$$\varepsilon(iv) = 1 + (n^2 - 1)/[1 + (v/v_e)^2], \qquad (6.33)$$

where n is the refractive index, roughly equal to $\sqrt{\varepsilon(v_{vis})}$. Substituting the above into Eq. (6.31) and integrating as before we obtain

$$w(r)_{v>0} = -\frac{\sqrt{3}hv_e a_1^3 a_2^3}{2r^6}$$

$$\times \frac{(n_1^2 - n_3^2)(n_2^2 - n_3^2)}{(n_1^2 + 2n_3^2)^{1/2}(n_2^2 + 2n_3^2)^{1/2}[(n_1^2 + 2n_3^2)^{1/2} + (n_2^2 + 2n_3^2)^{1/2}]}, \qquad (6.34)$$

where for simplicity, it is assumed that all three media have the same absorption frequency v_e. The total van der Waals interaction free energy of two identical molecules 1 in medium 3 (Fig. 13b) is therefore

$$w(r) = w(r)_{v=0} + w(r)_{v>0}$$

$$\approx -\left[3kT\left(\frac{\varepsilon_1(0) - \varepsilon_3(0)}{\varepsilon_1(0) + 2\varepsilon_3(0)}\right)^2 + \frac{\sqrt{3}hv_e}{4}\frac{(n_1^2 - n_3^2)^2}{(n_1^2 + 2n_3^2)^{3/2}} \right]\frac{a_1^6}{r^6}, \qquad (6.35)$$

which is strictly valid only for $r \gg a_1$.

The above equations highlight a number of important aspects of van der Waals forces in a solvent medium:

(i)   Since $hv_e \gg kT$ we find (as for interactions in free space) that the $v > 0$ dispersion force contribution is usually greater than the $v = 0$ dipolar contribution.

(ii)   The van der Waals force is much reduced in a solvent medium. For example, for two nonpolar molecules of refractive index $n_1 = 1.5$, the strength of the dispersion force in a solvent of $n_3 = 1.4$ will be reduced from its value in free space (where $n_3 = 1$) by a factor of

$$\frac{(1.5^2 - 1)^2(1.5^2 + 2)^{-3/2}}{(1.5^2 - 1.4^2)^2(1.5^2 + 2 \times 1.4^2)^{-3/2}} \approx 32.$$

(iii)   It is worth comparing Eq. (6.35) for the $v > 0$ dispersion contribution with London's equation for the free space interaction. If we put $n_3 = 1$, we obtain

$$w(r)_{v>0} = -\frac{\sqrt{3}hv_e a_1^6}{4r^6}\frac{(n_1^2 - 1)^2}{(n_1^2 + 2)^{3/2}} \qquad (6.36)$$

whereas London's result is

$$w(r) = -\frac{3hv_1\alpha_0^2}{4(4\pi\varepsilon_0)^2 r^6} = -\frac{3hv_1 a_1^6}{4r^6}\left(\frac{n_1^2 - 1}{n_1^2 + 2}\right)^2. \qquad (6.37)$$

The reason for the apparent discrepancy is that the absorption frequency of an isolated molecule v_I is different from that of the condensed phase v_e. It is a simple matter to ascertain that if $\alpha(v)$ and $\varepsilon(v)$ are related by the Clausius–Mossotti–Lorenz–Lorentz equation, $\alpha(v) \propto [\varepsilon(v) - 1]/[\varepsilon(v) + 2]$, then the two absorption frequencies are formally related by

$$v_e = \sqrt{3/(n_1^2 + 2)}\, v_I \tag{6.38}$$

so that when this is put into Eq. (6.36) we obtain the result based on London's formula, Eq. (6.37).

(iv)   The dispersion force between dissimilar molecules can be attractive or *repulsive* [if n_3 is intermediate between n_1 and n_2 in Eq. (6.34)]. Examples of repulsive van der Waals forces will be given in Chapter 11. The interaction between identical molecules, Eq. (6.35), is however always attractive, and we may further note that if the solute molecules 1 and solvent molecules 3 are interchanged (Fig. 13b→13c) the expression for the van der Waals force between the two solute (ex-solvent) molecules remains practically the same.

Equation (6.35) provides a semiquantitative criterion for determining which liquids are likely to be miscible and which are not. As a rule of thumb it is known that "like dissolves like." This is already apparent from Eq. (6.35) where we see that the smaller the difference between n_1 and n_3, the smaller the attraction between two solute molecules (Fig. 13b,c) and the less will be their tendency to associate (i.e., separate out into a different phase). Since Eq. (6.35) basically depends on the magnitude of $(n_1^2 - n_3^2)^2$ we may expect two liquids to become immiscible once $(n_1^2 - n_3^2)^2$ becomes too large. Now this can be written as $[\sqrt{(n_1^2 - 1)^2} - \sqrt{(n_3^2 - 1)^2}]^2$, which from the Lorenz–Lorentz equation, Eq. (5.29), is roughly proportional to $[\sqrt{\alpha_{01}^2/\sigma_1^6} - \sqrt{\alpha_{03}^2/\sigma_3^6}]^2$ or

$$w(r) \propto [\sqrt{U_1} - \sqrt{U_3}]^2, \tag{6.39}$$

where U_1 and U_3 are the cohesive energies (e.g., of vapourization) of the two liquids [cf. Eq. (6.5)]. This semiquantitative derivation forms the basis of Hildebrand's "solvent solubility parameter" δ, which is equal to the square root of the cohesive energy density of a liquid (Small, 1953). It is found that if two nonpolar liquids have similar values for δ [i.e., if $(\delta_1 - \delta_2)^2$ is small], they are miscible, and the binary mixture is nearly ideal (i.e., showing little deviation from Raoult's law) (for a review of solubility relationships see Kumar and Prausnitz, 1975).

(v)   In some instances the dispersion force between two molecules in a solvent is very small, and then the zero-frequency contribution dominates the interaction. The best-known case concerns the lower molecular weight alkanes whose refractive indices are very close to that of water (cf. $n_{CH_4} \approx$ 1.30, $n_{C_4H_{10}} \approx 1.33$, $n_{C_5H_{12}} \approx 1.36$, with $n_{H_2O} \approx 1.33$). When these molecules interact in water (or vice versa) the major contribution is now not the dispersion force but rather the first term in Eq. (6.35), and since $\varepsilon_{H_2O}(0) \approx 80$ while $\varepsilon_{\text{alkane}}(0) \approx 2$, this reduces to approximately

$$w(r)_{v=0} \approx -\frac{kTa_1^6}{r^6}, \tag{6.40}$$

which is purely entropic, and indicates that there is an increase in entropy (greater motional freedom of water molecules) as two alkane molecules approach each other. It has long been known that the "hydrophobic interaction" between small hydrocarbon molecules in water is mainly entropic, but the few measured values to date suggest a far stronger interaction than would be expected from Eq. (6.40). Thus, for two small molecules of radius a, we would expect a free energy of dimerization of order $-kT \, (a/2a)^6 \approx -kT/64$, or about 0.04 kJ mol^{-1} at 25°C (not enough to induce immiscibility), whereas the experimental values are at least 100 times larger, viz., ~ 10 kJ mol^{-1} for CH_4, C_6H_6, and C_6H_{12} (Tucker et al., 1981; Ben Naim et al., 1973). This lack of agreement is related to the breakdown in the simple model of the excess polarizability for highly polar molecules, as discussed in Section 5.7 (and Table VII). The unique and unusual properties of water, both as a solvent and as a suspending medium for solute–solute interactions, are further investigated in Chapter 8 and in Part II.

6.7. Dispersion self-energy of a molecule in a medium

The concept of a dispersion self-energy of a molecule is analogous to the Born self-energy of an ion (Chapter 3), and, as for ions, it provides insight into the *solubility* and *partitioning* of molecules in different solvents. There are a number of different approaches to this problem (Mahanty and Ninham, 1976); we shall adopt the simplest, which nevertheless brings out all the essential physics. Let us consider the transfer of a molecule 1 of diameter σ_1 from free space into a medium 2 (Fig. 13d), where it becomes surrounded by 12 solvent molecules of similar diameter σ_2. The free energy change is therefore given by the (positive) energy needed to first create a cavity, which involves breaking six solvent–solvent "bonds," plus the (negative) energy of placing the molecule in the cavity, which involves the formation of 12 new solute–solvent "bonds." Thus, for this process,

$$\Delta\mu_{\text{disp}}^i \approx \frac{3h\nu_1}{4(4\pi\varepsilon_0)^2}\left[\frac{6\alpha_{02}^2}{\sigma_2^6} - \frac{12\alpha_{01}\alpha_{02}}{\sigma_1^3\sigma_2^3}\right], \tag{6.41}$$

where ν_1 is assumed to be the same for the solute and solvent molecules. We may already note that if medium 2 = medium 1 (Fig. 13e), the above reduces to the result based on the London equation, Eq. (6.5), which gives the cohesive energy (latent heat) of the pure liquid or solid medium.

More generally, the free energy of transfer of a molecule 1 from medium 3 into medium 2 (Fig. 13f) is therefore given by

$$\Delta\mu_{\text{disp}}^i \approx \frac{3h\nu_1}{4(4\pi\varepsilon_0)^2}\left[-\left(\frac{6\alpha_{03}^2}{\sigma_3^6} - \frac{12\alpha_{01}\alpha_{03}}{\sigma_1^3\sigma_3^3}\right) + \left(\frac{6\alpha_{02}^2}{\sigma_2^6} - \frac{12\alpha_{01}\alpha_{02}}{\sigma_1^3\sigma_2^3}\right)\right], \quad (6.42)$$

which, using Eq. (5.29), is roughly proportional to

$$\Delta\mu_{\text{disp}}^i \propto (\alpha_{03} - \alpha_{02})(2\alpha_{01} - \alpha_{02} - \alpha_{03})$$

$$\propto (n_3^2 - n_2^2)(2n_1^2 - n_2^2 - n_3^2), \quad (6.43)$$

which can be positive or negative. In particular, the above shows that solute transfer is always energetically favoured ($\Delta\mu^i$ negative) into the solvent whose refractive index is closer to that of the solute molecule—a further manifestation of the "like-dissolves-like" rule. If medium 2 = medium 1 (Fig. 13g), the above reduces to

$$\Delta\mu_{\text{disp}}^i \approx -\frac{6(3h\nu_1)}{4(4\pi\varepsilon_0)^2}\left[\frac{\alpha_{01}}{\sigma_1^3} - \frac{\alpha_{03}}{\sigma_3^3}\right]^2 \quad (6.44)$$

$$\propto -(\sqrt{U_1} - \sqrt{U_3})^2 \propto -(n_1^2 - n_3^2)^2, \quad (6.45)$$

which is always negative and which applies equally to the two transfer processes shown in Fig. 13g. Thus, it is always *energetically* favourable for a solute molecule to move into its own environment. It is interesting to note that the above is essentially the same result that was obtained in the previous section, Eq. (6.39), and leads to the same semiempirical solubility relations as contained in the "solubility parameter" criterion. In particular, the very low solubility of alkanes and hydrocarbons in water, where $(n_1^2 - n_3^2) \approx 0$, is once again seen not to be governed by their dispersion interaction, but by other factors.

The above relationships and rule-of-thumb criteria, while useful, are limited to dispersion interactions and thus require that the solutes and solvents be nonpolar or only weakly polar. In Chapter 9 we shall develop similar relationships, based on cohesive energies, which can be applied to other phenomena as well (e.g., wetting).

6.8. Further aspects of van der Waals forces

(i) *Anisotropy of dispersion forces.* The polarizabilities of all but spherically symmetric molecules are anisotropic, having different values along

different molecular directions. This arises because the electronic polarizabilities of bonds (which measure the response of the electrons to an external field) are usually anisotropic (Hirschfelder *et al.*, 1954). For example,
the longitudinal and transverse polarizabilities of the C—H bond are
$\alpha_{\parallel}/4\pi\varepsilon_0 = 0.79 \times 10^{-30}$ m^3 and $\alpha_{\perp}/4\pi\varepsilon_0 = 0.58 \times 10^{-30}$ m^3; the mean
value (quoted in Table VI) is $\frac{1}{3}(\alpha_{\parallel} + 2\alpha_{\perp})/4\pi\varepsilon_0 = 0.65 \times 10^{-30}$ m^3. One consequence of the anisotropy in α is that the dispersion force between mole-
cules becomes dependent on their mutual orientation (see Hirschfelder *et al.*,
1954; Israelachvili and Tabor, 1973; Israelachvili, 1974). In nonpolar liquids
the effect is not important, since the molecules are tumbling rapidly, and
their *mean* polarizability is then what matters; but in solids and liquid
crystals the anisotropic attractive forces can sometimes be an important
factor in driving molecules into favourable mutual orientations; for example, in determining the specific configurations of macromolecules such as
polymers and proteins, the ordering of molecules in liquid crystals, and the
temperature-dependent phase transitions of lipid bilayers and monolayers
(Zwanzig, 1963; Marčelja, 1973).

It should be remembered, however, that for asymmetric molecules, the
repulsive forces are also orientation-dependent (reflecting their asymmetric
shape), and this is usually far more important in determining how molecules
align themselves relative to each other (see Chapter 7).

(ii)   *Nonadditivity of van der Waals forces and many-body effects.* Unlike
gravitational and Coulomb forces, van der Waals forces are not generally
pair-wise additive; the force between any two molecules is affected by the
presence of other molecules nearby, so that one cannot simply add all the pair
potentials of a molecule to obtain its net interaction energy with all the other
molecules. This is because the field emanating from any one molecule reaches a
second molecule both directly and by "reflection" from other molecules (since
they, too, are polarized by the field). The net effect is that the van der Waals
force between two molecules in a medium is "enhanced" and therefore larger
than would be expected from a consideration of two-body (pair) forces alone.
The effect is small (up to 30%) but can be important; for example, the total
energy contributions, including *many-body* effects, to the lattice energies of the
rare gas atoms (such as argon and xenon) determine that they pack as face-
centered cubic (FCC) solids and not as hexagonal close-packed (HCP) solids.
The nonadditive property of van der Waals forces is particularly important in
the interactions between large particles and surfaces (Chapter 11).

(iii)   *Retardation effects.* When two atoms are an appreciable distance
apart, the time taken for the electric field of the first atom to reach the second
and return can become comparable with the period of the fluctuating dipole
itself (cf. the distance travelled by light during one rotation of a Bohr atom
electron is about 3×10^8 m s$^{-1}/3 \times 10^{15}$ s$^{-1} \approx 10^{-7}$ m or 100 nm). When this

happens the field returns to find that the direction of the instantaneous dipole of the first atom is now different from the original and less favourably disposed to an attractive interaction. Thus, with increasing separation the dispersion energy between two atoms begins to decay even faster than $1/r^6$. This is called the *retardation effect*, and the dispersion forces between molecules and particles at large separations are called *retarded forces*.

For two molecules in free space, retardation effects begin at separations of about 5 nm and are therefore of little interest. However, in a medium, where the speed of light is slower, retardation effects come in at smaller distances, and they become particularly important when macroscopic bodies or surfaces interact in a liquid medium as we shall see in later chapters. Note that it is only the dispersion energy that suffers retardation; the zero frequency orientation and induction energies remain nonretarded at all separations, so that as the separation increases these eventually dominate the van der Waals interaction.

Chapter 7

Repulsive Forces, Total Intermolecular Pair Potentials, and Liquid Structure

7.1. Sizes of atoms, molecules, and ions

At very small interatomic distances the electron clouds of atoms overlap, and there arises a strong repulsive force that determines how close two atoms or molecules can ultimately approach each other. These repulsive forces are sometimes referred to as *exchange repulsion, hard core repulsion, steric repulsion*, or (for ions) *Born repulsion*, and they are characterised by having very short range and increasing very sharply as two molecules come together. Strictly, they belong to the category of quantum mechanical or chemical forces discussed earlier, and unfortunately there is no general equation for describing their distance dependence. Instead, a number of empirical potential functions have been introduced over the years, all of which appear to be satisfactory so long as they have the property of a steeply rising repulsion at small separations. The three most common of such potentials are the *hard sphere potential*, the inverse *power-law potential*, and the *exponential potential*. Let us begin by considering the first.

If atoms are considered as *hard spheres*, (i.e., incompressible), the repulsive force suddenly becomes infinite at a certain interatomic separation. This simple model reflects the observation that when different atoms pack together in liquids and solids they often do behave as hard spheres, or "billiard balls," of fixed radii characteristic for each atom. Defined in this way the radius of an atom (whether isolated or covalently bound) or a spherical molecule is then called its *hard sphere radius* or *van der Waals packing radius*. Results obtained from x-ray and neutron diffraction studies on solids (especially crystals) and liquids yield values that agree to within a few percentage points. The van der Waals packing radii of most atoms and small molecules lie between 0.1 and 0.2 nm, some of which are illustrated in Fig. 14. Similar concepts apply to ions in ionic crystals, where the characteristic packing radius is referred to as the *bare ion radius*. The bare ion radius is quite different from the hydrated ion radius of an ion in water, discussed in Section 4.5. Some bare ion radii are also shown in Fig. 14.

It is also recognized that the distance between the atomic centres of two atoms connected by a covalent bond can be expressed as the sum of their *covalent bond radii* (Fig. 14). Single-bond covalent radii are usually about 0.08 nm shorter than the nonbonded (van der Waals) radii.

By considering both the covalent and van der Waals radii of individual atoms in a molecule and the bond angles between them, one can establish the effective van der Waals radius of a molecule. This concept is strictly valid only for small, nearly spherical molecules such as CH_4, but Bondi (1968) described a procedure for calculating the effective van der Waals radii of molecules and molecular groups, some of which are included in Fig. 14. For very asymmetric molecules that cannot be considered even as quasi-spheres, one has to consider their *van der Waals dimensions* along different molecular axes. For example, in alkanes the effective radius of the cylinder-like paraffin chain is about 0.20 nm, while its total length is given by ~ 0.127 nm per CH_2-CH_2 link along the chain plus ~ 0.20 nm for each hemispherical CH_3 group at each end.

There are many different methods for measuring the radii of atoms and molecules. These include $P-V-T$ and spectroscopic data on gases; viscosity and diffusion data on both gases and liquids; and compressibility, electron, x-ray, and neutron diffraction data on liquids and solids. The values for molecular sizes deduced from these different methods can differ by as much as 30%, as illustrated in Table XI. This is because each method measures a slightly different property. Thus, molecular radii $\sigma/2$ determined from the van der Waals equation of state coefficient b, Eq. (6.13),

$$\sigma/2 = 0.463b^{1/3} \quad \text{nm} \quad (b \text{ in } dm^3 \text{ mol}^{-1}) \tag{7.1}$$

gives the smallest value since it reflects the sizes of molecules during collisions, when they approach each other closer than their equilibrium separation. Values determined from the viscosity and self-diffusion of molecules in liquids (Dymond, 1981; Evans *et al.*, 1981) usually yield results similar to those obtained from crystal packing (i.e., the van der Waals radii) while radii calculated from the molecular volume occupied in the liquid state (even assuming close-packing),

$$\tfrac{4}{3}\pi(\sigma/2)^3 = 0.7405(M/N_0\rho), \tag{7.2}$$

	Li^+	Na^+	K^+	NH_4^+	Cs^+	$N(CH_3)_4^+$
Monovalent cations	0.068	0.095	0.133	0.148	0.169	0.347

	Be^{2+}	Mg^{2+}	Fe^{2+}	Ca^{2+}	Ba^{2+}	
Divalent cations	0.031	0.065	0.076	0.099	0.135	

	Al^{3+}	Fe^{3+}	La^{3+}			
Trivalent cations	0.050	0.064	0.104			

	F^-	OH^-	Cl^-	Br^-	I^-	NO_3^-
Monovalent anions	0.136	0.176	0.181	0.195	0.216	0.264

	Ne	Ar	Kr	Xe	CH_4	CCl_4
Spherical molecules	0.154	0.188	0.201	0.216	0.20	0.275

Effective radii of nonspherical molecules and groups (approximate)	H_2O 0.14	O_2 0.18	NH_3 0.18	HCl 0.18	HBr 0.19
	CH_3OH 0.21	CH_3Cl 0.215	$CHCl_3$ 0.255	C_6H_6 0.265	C_6H_{12} 0.285
	$-CH_3$ group 0.20	$-CH_2-$ group 0.20	$-NH_2$ group 0.17	$-OH$ group 0.145	Aromatic ring thickness 0.37

Exposed radii of atoms covalently bonded in molecules 	H 0.11	F 0.14	O 0.15	N 0.15	C 0.17
	Cl 0.18	S 0.18	Br 0.19	P 0.19	I 0.20

Covalent bond radii of atoms 	$-H$ 0.03	$-C$ 0.077	$-N$ 0.070	$-O$ 0.066	$-F$ 0.064	$-S$ 0.104
		$=C$ 0.067	$=N$ 0.062	$=O$ 0.062	$-Cl$ 0.099	$-P$ 0.110
		$\equiv C$ 0.060	$\equiv N$ 0.055		$-Br$ 0.114	$-Si$ 0.117

Fig. 14 Effective packing radii of atoms, molecules and ions (nm).

TABLE XI
Radii of molecules deduced from different methods

Molecule	Minimum Radius (nm) From van der Waals Coefficient b,[a] Eq. (7.1)	Mean Radius (nm) van der Waals Packing Radius (From Fig. 14)	Maximum Radius (nm) From Volume Occupied in Liquid at $20°C$, Eq. (7.2)
CH_3OH	0.19	0.21	0.23
$CHCl_3$	0.22	0.26	0.29
C_6H_6	0.23	0.27	0.30
CCl_4	0.24	0.28	0.305

[a] Constants b taken from the "Handbook of Chemistry and Physics," CRC Press, 56th ed.

yield the highest values because in the liquid the mobile molecules are *on average* about 5% farther apart than in the close-packed crystalline solid.

7.2. Repulsive potentials

Returning now to our discussion of repulsive potentials, the *hard sphere potential* can be described by

$$w(r) = +(\sigma/r)^n, \qquad n = \infty. \tag{7.3}$$

Since for $r > \sigma$ the value of $w(r)$ is effectively zero while for $r < \sigma$ it is infinite, this expression nicely describes the hard sphere repulsion where σ is the *hard sphere diameter* of an atom or molecule, which may be associated with twice the van der Waals radius.

Two other repulsive potentials are worthy of note: the *power-law potential*

$$w(r) = +(\sigma/r)^n, \tag{7.4}$$

where n is now an integer (usually taken between 9 and 16), and the *exponential potential*

$$w(r) = +ce^{-r/\sigma_0}, \tag{7.5}$$

where c and σ_0 are adjustable constants, with σ_0 usually varying between 0.02 and 0.03 nm. Both of these potentials are more realistic in that they allow for the finite compressibility or "softness" of atoms. The power-law potential has

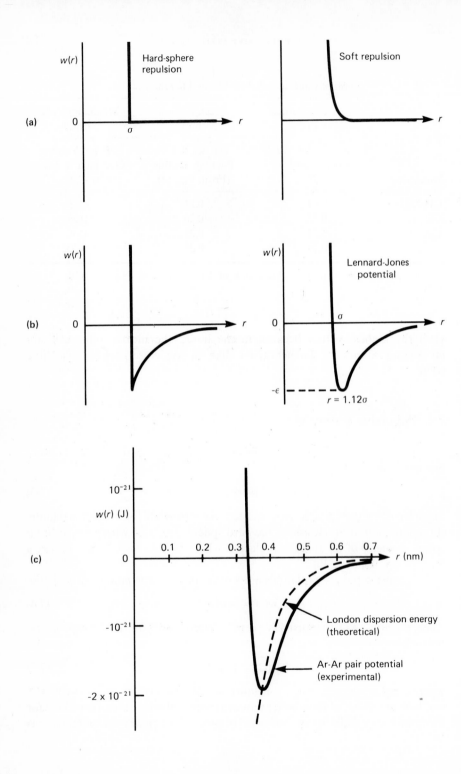

little theoretical basis, while the exponential potential has some theoretical justification; it is generally recognized, however, that their common usage is due mainly to their mathematical convenience.

7.3. Total intermolecular pair potentials

The *total intermolecular pair potential* is obtained by summing the attractive and repulsive potentials. Figure 15a illustrates the two main types of repulsive potentials discussed above and the shapes of the potential functions they lead to when a long-range inverse power attractive term is added (Fig. 15b). The best known of these is the *Lennard–Jones* or *"6–12" potential*:

$$w(r) = c_1/r^{12} - c_2/r^6 = 4\varepsilon[(\sigma/r)^{12} - (\sigma/r)^6], \qquad (7.6)$$

which is widely used because of its simplicity and inverse sixth-power attractive van der Waals term. The particular mathematical form given for the Lennard–Jones potential above is commonly used though it is well to note that the parameter σ is different from the molecular diameter. Thus, for the Lennard–Jones potential, $w(r) = 0$ at $r = \sigma$, and the minimum energy occurs at $r = 2^{1/6}\sigma = 1.12\sigma$ (Fig. 15b). The minimum energy is $w(r) = -\varepsilon$, where the attractive van der Waals contribution is -2ε while the repulsive energy contribution is $+\varepsilon$. Thus the inverse 12th-power repulsive term decreases the strength of the binding energy at equilibrium by 50% (from -2ε to $-\varepsilon$). This may be compared to the hard sphere repulsion potential where the binding energy at equilibrium would be just the van der Waals energy at contact (at $r = \sigma$). It is for this reason that the lattice energies of ionic crystals and van der Waals solids, calculated on the basis of a hard sphere repulsion, are sometimes too large (cf. Section 3.5). But quite often the hard sphere potential together with a London dispersion attraction gives excellent results when compared with the experimental molar lattice energies of nonpolar solids (cf. Table VIII). This is because the *attractive* forces are usually *larger* than predicted by the simple London equation, which ignores contributions from other absorption frequencies, quadrupole interactions, and—in solids and liquids—many-body effects. On the other hand, the *repulsive* potentials are believed to be steeper (i.e., *weaker* at the equilibrium separation) than given by the $1/r^{12}$ term in the Lennard–Jones potential. All these effects conspire to make the hard

---

Fig. 15 (a) Repulsive potentials.

(b) Full pair potentials obtained by adding an attractive long-range potential to the above repulsive potentials. The packing diameter of a molecule is given by the value of r at the potential energy minimum.

(c) Experimental Ar–Ar potential compared with the London dispersion energy. (c) From Parsons *et al.*, 1972.)

sphere potential a reasonable one for predicting contact energies. Figure 15c nicely illustrates this fortuitous "cancellation of errors": here we see the experimentally determined pair potential for argon (Parsons et al., 1972) together with the London dispersion energy, $w(r) = -50 \times 10^{-79}/r^6$ J, taken from Table VIII. Note how the (theoretical) dispersion attraction passes through the experimental curve almost exactly at its minimum value.

7.4. Role of repulsive forces in non–covalently bonded solids

Let us now turn our attention to the role of repulsive forces in solids. More often than not it is these that determine many of the properties of solids. The reason for this is that for many types of attractive interactions, their orientation or angle dependence is weak (i.e., they are nondirectional). This is particularly so for van der Waals forces. In contrast, the orientation dependence of the repulsion, which simply reflects the asymmetric shape of a molecule, often has a large effect. Thus, when molecules or ions come together in the condensed state, the way they can pack together—reflected in their relative sizes and shapes—now becomes a major consideration in determining their lattice structures and melting points, but not their boiling points. To see why this is so, recall that latent heats of vapourization can be closely associated with the cohesive energies in solids and liquids (Table I). Thus it is the *attractive* forces that mainly determine latent heats of vapourization and, therefore, also of boiling points. Melting points, however, are determined by the geometry of molecules. If their shape allows them to comfortably pack together into a lattice, they will tend to remain in this state and will therefore have a high melting point. If their shape does not allow for good packing, the melting point will be low. (Note too that latent heats of melting are usually fairly small, an indication that on melting few "bonds" are broken and that the main effect is a rearrangement of molecular ordering.) Since the shape of a molecule is determined by its van der Waals dimensions, which is effectively a statement about its repulsive force, we can now see why *repulsive forces* are mainly responsible for melting while *attractive forces* are mainly responsible for boiling.

This somewhat broad generalization can be illustrated by the melting points and boiling points of some hydrocarbons, shown in Table XII. Here we clearly see that the T_B values are not much affected by branching, or by replacing a single C—C bond by a double bond, since the intermolecular attractive forces are not much changed. But the T_M values are lowered dramatically, since the regular crystalline packing of the all-*trans* chains is no longer possible when a "kink" is present somewhere along the otherwise cylindrical molecules. Note

TABLE XII

Effect of double bonds and chain branching on
melting and boiling points of hydrocarbons

Hydrocarbon		T_M (°C)	T_B (°C)
C_6H_{14}	n-Hexane	−95	69
C_6H_{12}	1-Hexene	−140	63
C_6H_{12}	2-Hexene (cis)	−141	69
C_6H_{14}	2-Methyl pentane	−154	60
C_8H_{18}	n-Octane	−57	126
C_8H_{16}	1-Octane	−102	121
C_8H_{16}	2-Octane (cis)	−100	126
C_8H_{18}	2-Methyl heptane	−120	119
$C_{12}H_{26}$	n-Dodecane	−10	216
$C_{12}H_{24}$	1-Dodecene	−35	213
$C_{16}H_{34}$	n-Hexadecane	18	287
$C_{16}H_{32}$	1-Hexadecene	4	284

how the effect of a single kink persists even for the 16-carbon hexadecane molecule.

For some molecules, however, the orientation dependence of their *attractive* forces plays the dominant role in determining their solid- and liquid-state properties. This is of course especially true of covalent bonds but also when strongly directional dipolar and hydrogen-bonding interactions are involved. A particularly notable representative of this phenomenon is the water molecule, whose interactions are so remarkable (and important) that they will be given special attention in the next chapter.

The sizes and shapes of surfactant and biological macromolecules such as lipids and proteins are particularly important in determining how they can pack (assemble) into aggregated structures such as micelles and membranes in water, as will be discussed in Part III.

7.5. Role of repulsive forces in liquids: liquid structure

When a solid melts, the ordered molecular structure that existed in the solid is not completely lost in the liquid. This phenomenon has led to such concepts as *liquid structure* and *molecular ordering* in liquids, with important consequences both for our understanding of the liquid state and for the way molecules and particles interact in liquids (Pryde, 1966; Kohler, 1972; Croxton, 1975; Kruus, 1977; Maitland *et al.*, 1981). The occurrence of "structure" in liquids arises first and foremost from the size and shape of molecules, and as such reflects the repulsive forces between them. Let us see what it is all about by first considering the molecular events that take place as a solid is heated through its melting point.

Imagine a close packed FCC lattice at 0 K where each molecule is surrounded by 12 nearest neighbours at a distance $r = \sigma$, six next-nearest neighbours at $r = \sqrt{2}\sigma$, 24 at $r = \sqrt{3}\sigma$, and so on. Figure 16a shows the number n of molecules to be found at a radial distance r away from any central or reference molecule. At a higher temperature, but still below the melting

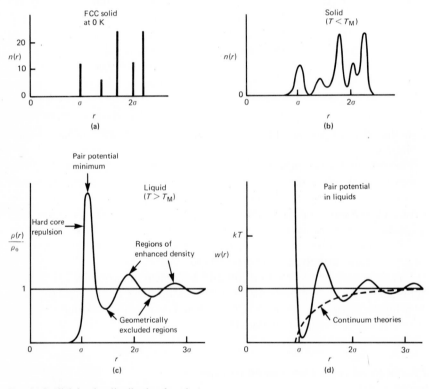

Fig. 16 Radial density distribution functions:
(a) for a close-packed FCC solid at 0 K;
(b) for a solid at a finite temperature below its melting point T_M;
(c) for a liquid; such data is obtained from x-ray and neutron scattering experiments on liquids (Pryde, 1966; Kohler, 1972; Kruus, 1977).
(d) Pair potential for two molecules in a liquid whose density distribution function is shown on the left, computed from $\rho(r) = \rho_0 \exp[-w(r)/kT]$.

point, the mean distance between the molecules increases slightly and the amplitude of molecular vibrations also increases; but the ordered lattice structure is maintained. The number density of molecules around the reference molecule now looks like Fig. 16b, and since it is now spread out in space rather than a set of discrete lines one can no longer talk of the number n at r, but more of the probability density of finding a molecule at r.

At the melting point the ordered lattice structure breaks down abruptly. This occurs once the amplitude of the molecular vibrations reaches a value where the molecules can move out of their confined lattice sites (i.e., they can now move past each other). Theoretically, for hard spheres, this ability should occur at a density of about 0.86 of the close-packed (van der Waals) solid, corresponding to a volume expansion of about 16%. This is borne out by experiments where on melting, the increased volume is usually of this order, especially for spherical molecules (cf. 15–16% for Ne, Ar, Kr, Xe, and CH_4, 17.5% for n-hexane, 20% for benzene). But it can be much larger (e.g., 31% for polyethylene) or much smaller, and even negative as for water (Bondi, 1968).

But even in the liquid state the molecules are still very much restricted in their motion and in the way they can position themselves with respect to each other. The 16% increase in space available is not much, and the tendency to pack into an ordered lattice persists in the liquid. Thus, our spherical reference molecule will now have only slightly fewer nearest neighbours, about 8 to 11 instead of 12, which will tend to group around it in a nonrandom way. The next-nearest neighbours will likewise order around the first group but with a smaller degree of *correlation*. Eventually, at larger distances, there will be no correlation at all with respect to our reference molecule. This short-range order extends over a few molecular diameters and characterises the liquid state (in the crystalline solid the order extends indefinitely).

Thus, radially away from any reference molecule the liquid density profile $\rho(r)$ looks like Fig. 16c. The density profile is usually plotted as $\rho(r)/\rho_0$ against r, where ρ_0 is the density of the bulk liquid. $\rho(r)/\rho_0$ is commonly referred to as the *radial distribution function* or the *pair correlation function*. At large distances it always approaches unity as $\rho(r)$ approaches ρ_0.

The magnitude and range of molecular ordering in liquids is enhanced by increasing the external pressure and lowering the temperature, as may be expected. Rigid hard sphere molecules exhibit more short-range order or structure than deformable molecules, such as n-hexane, where internal bond flexibility allows for greater configurational freedom to pack in different ways. The net effect, however, is that the density $\rho(r)$ usually oscillates with distance away from any molecule and only reaches the bulk liquid value ρ_0 at some distance away. Thus, in general a liquid medium near a molecule or surface will not have the properties of a structureless continuum (for which the density would be equal to the bulk value right up to contact).

7.6. Effect of liquid structure on molecular forces

For two solute molecules or particles 1 dissolved in a liquid medium 2, the problem becomes exceedingly complicated. Three pair correlation functions may now be identified: $\rho_{11}(r)$ for the *solute–solute* density profile [i.e., the variation of $\rho_1(r)$ away from a solute molecule within the solvent medium], $\rho_{22}(r)$ for the *solvent–solvent* density profile, and $\rho_{12}(r)$ for the *solute–solvent* density profile. Each of these will exhibit oscillatory behaviours that are interdependent, depending on the sizes of the molecules and on the nature of the solute–solute, solute–solvent, and solvent–solvent interaction forces. For example, solute–solvent interactions will affect the ordering of solvent molecules around a solute molecule, which in turn affects the solute density profile around the solute molecule.

How does all this affect the interaction between two dissolved molecules? In Chapter 2 we saw that the effective pair potential (or potential of mean force) between two solute molecules at a distance r apart is related to their density $\rho(r)$ at r by the Boltzmann relation:

$$\rho_{11}(r) = \rho_0 \exp[-w_{11}(r)/kT]$$

or

(7.7)

$$\rho_{11}(r)/\rho_0 = \exp[-w_{11}(r)/kT].$$

Thus, $w_{11}(r)$ tends to zero at large r, where $\rho(r)/\rho_0 \to 1$, but *oscillates with distance* at smaller r; $w_{22}(r)$ behaves similarly. Such a pair potential is shown schematically in Fig. 16d. We see therefore how liquid structure can dramatically influence the interaction between dissolved molecules already at large distances. Unfortunately there is a lack of any direct experimental data concerning the effective pair potentials of solute molecules and small particles interacting in a (different) solvent medium and, in particular, how reliable continuum theories are in determining the first potential-energy minimum (Fig. 16c). But this is not the case for interacting surfaces, and in Chapter 13 we shall see how solvent structure arises and affects forces between macroscopic surfaces at distances below a few molecular diameters, for which both theoretical and experimental data are now available.

Chapter 8

Special Interactions: Hydrogen-Bonding, Hydrophobic, and Hydrophilic Interactions

8.1. The unique properties of water

Water is such an unusual substance that it has been accorded a special place in the annals of phenomena dealing with intermolecular forces, and two types of "special interactions"—the *hydrogen bond* and the *hydrophobic effect*—are particularly relevant to the interactions of water. The literature on the subject is vast (Franks, 1972–1982), not only because water is the most important liquid on earth, but also because it has so many interesting and anomalous properties.

For a liquid of such a low molecular weight, water has unexpectedly high melting and boiling points and latent heat of vapourization. There are, of course, many substances of low molecular weight and high melting and boiling points, but these are invariably ionic crystals or metals whose atoms are held together by strong Coulombic or metallic bonds. Thus, these properties of water point to the existence of an intermolecular interaction that is stronger than that expected for ordinary, even highly polar, liquids.

The density maximum at 4°C exhibited by liquid water, and the unusual phenomenon that the solid (ice) is lighter than the liquid, indicates that in the ice lattice the molecules prefer to be farther apart than in the liquid. Thus, we may further conclude that the strong intermolecular bonds formed in ice persist into the liquid state and that they must be strongly orientation-dependent since water adopts a tetrahedral coordination (four nearest neighbours per molecule) rather than a higher packing density (cf. 12 nearest neighbours characteristic of close-packed van der Waals solids where the "bonds" are nondirectional). Water has other unusual properties, such as a high molecular dipole moment and dielectric constant, and very low compressibility, but enough has already been said to indicate that water is a very complex liquid and that some unusually strong and orientation-dependent bonds are involved.

8.2. The hydrogen bond

It is a straightforward matter to ascertain which bond is responsible for this interaction by simply looking at the distances between various atomic centers in the ice lattice (Fig. 17a). The *intramolecular* O—H distance is about 0.10 nm, as expected for this covalent bond (see Fig. 14), but the *intermolecular* O$\cdots$H distance is only 0.176 nm, much less than the 0.26 nm expected from summing the two van der Waals radii but still larger than the covalent distance of 0.10 nm. Thus, the intermolecular O$\cdots$H bond is implicated, which at first sight appears to possess some covalent character. Such bonds are known as *hydrogen bonds*, and the reader is referred to Schuster *et al.* (1976) and Joesten and Schaad (1974) for the voluminous literature on the subject.

Hydrogen bonds are not unique to water; they exist to varying degrees between electronegative atoms (e.g., O, N, F, and Cl) and H atoms covalently bound to similar electronegative atoms. These bonds are special in that they only involve hydrogen atoms, which, by virtue of their tendency to become positively polarized and their uniquely small size, can interact strongly with nearby electronegative atoms resulting in an effective H-mediated "bond" between two electronegative atoms. Originally, it was believed that the hydrogen bond was quasi-covalent and that it involved the sharing of a H atom or proton between two electronegative atoms. But it is now accepted (Coulson, 1961; Umeyama and Morokuma, 1977) that the hydrogen bond is predominantly an electrostatic interaction: with few exceptions, the H atom is not shared but remains closer to and covalently bound to its parent atom. The hydrogen bond between two groups XH and Y is usually denoted by X—H$\cdots$Y.

The strengths of most hydrogen bonds lie between 10 and 40 kJ mol^{-1} (Joesten and Schaad, 1974), which makes them stronger than a typical van der Waals "bond" (~ 1 kJ mol^{-1}) but still much weaker than covalent bonds (~ 500 kJ mol^{-1}).

Hydrogen bonds can occur intermolecularly as well as intramolecularly and can happily exist in a nonpolar environment. They are consequently particularly important in macromolecular and biological assemblies, such as in proteins, linking different segments together inside the molecules, and in nucleic acids, where they are responsible for the stability of the DNA molecule (holding the two helical strands together). Their involvement in setting up

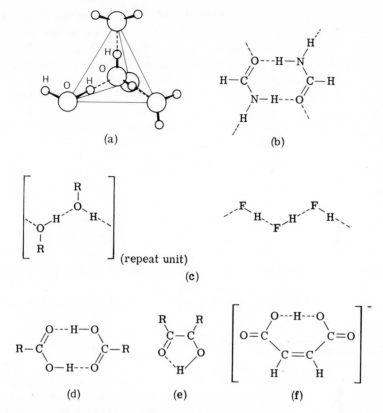

Fig. 17 Different types of hydrogen bonds and hydrogen-bonded structures. Linear hydrogen bonds have the lowest energy, but some H bonds with a —H--- angle of 120° or less also occur.
(a) Three-dimensional structures (e.g., ice).
(b) two-dimensional (layered) structures (e.g., formamide).
(c) one-dimensional (chain and ring) structures (e.g., alcohols, HF).
(d) dimers (e.g., fatty acids).
(e) intramolecular H bond.
(f) symmetric H bond (H atom shared).

one-, two-, and three-dimensional macromolecular structures is sometimes referred to as *hydrogen-bond polymerization* (illustrated in Fig. 17).

8.3. Models of water and associated liquids

Hydrogen bonds play a particularly prominent role in water since each oxygen atom with its two hydrogens can participate in four such linkages with other water molecules—two involving its own H atoms and two involving its unshared (lone-pair) electrons with foreign H atoms. To see exactly how this arises we require some picture of the charge distribution within the water molecule. Various models have been proposed of which the so called ST2 model of water [named after a modified model by Stillinger and Rahman (1974)] will be described. Other models also exist (Finney, 1982), and it is recognized that as yet no single model has been able to satisfactorily account for the properties of water in all three phases (ice, liquid, and vapour). However, the main features of ST2 water are simple and quite similar to some of the other models and so provide a good introduction to the various conceptual approaches that are being applied in attempts at understanding water.

In ST2 water (Fig. 18) the water molecule is modelled with charges of $+0.24e$ centered on each hydrogen atom and two compensating charges of $-0.24e$ on the opposite side of the oxygen atom, representing the two unshared electron pairs. The four charges are located along four tetrahedral arms radiating out from the center of the O atom. The interaction between any two water molecules is now assumed to involve an isotropic Lennard–Jones potential and 16 Coulombic terms representing the interaction between each of the four point charges on one molecule with the four on the other. The net Coulombic interaction obviously depends on the mutual orientation of the two molecules. When many molecules are involved their equilibrium configuration can only be solved on a computer, and when this is done the model can account for a number of the unusual properties of water, such as the highly open ice structure and the density maximum in the liquid state. Computer simulations show that this comes about because of the strong preference for the molecules to adopt a lattice where each oxygen is tetrahedrally coordinated to four other oxygens, with each hydrogen atom

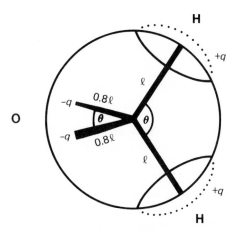

Fig. 18 ST2 model of water molecule; $q = 0.24e$, $l = 0.1$ nm, $\theta = 109°$.

lying in the line joining two oxygen atoms. It is this preferred linearity of the O—H $\cdots$ O bond in water that endows it with its strongly directional nature.

In liquid water the tendency to retain the ice-like tetrahedral network remains, but the structure is now disordered and labile. The average number of nearest neighbours per molecule *rises* to about five (hence the higher density of water on melting), but the mean number of H bonds per molecule *falls* to about three whose lifetimes are estimated to be about 10^{-11} s. Other strongly hydrogen-bonding molecules, such as formamide and HF, also retain some of their ordered crystalline structure in the liquid state over short distances; such liquids are known as *associated liquids*. It is also believed that the H-bond structure in such liquids is *cooperative* in the sense that the presence of H bonds between some molecules enhance their formation in nearby molecules, thereby tending to propagate the H-bonded network. If so, the interaction is non–pairwise additive, which presents serious problems in theoretical computations of H-bond association.

Finally, it is well to note that the tetrahedral coordination of the water molecule is at the heart of the unusual properties of water, much more than the hydrogen bonds that bring them about. As a rule of thumb, molecules that can participate in only two H bonds can link up into a one-dimensional chain or ring (e.g., HF and alcohols) (Fig. 17). Likewise, atoms of valence two such as selenium and tellurium can form long chains of covalently bonded atoms. Atoms that can participate in three bonds [e.g., arsenic, antimony, and carbon (in graphite)], can form two-dimensional sheets or layered structures held together by weaker van der Waals forces. But only a tetrahedral, or higher, coordination allows for a three-dimensional network to form. For example, it is the tetrahedral coordination characteristic of carbon and silicon that results in their almost infinite variety of atomic associations whether in chain molecules (e.g., polymers, surfactants, polypeptides), cyclic compounds, or two- and three-dimensional crystals (e.g., diamond, silica, sheet silicates).

8.4. Relative strengths of different types of interactions

In earlier chapters we saw how for molecules where only one type of attractive force is present (e.g., pure Coulombic or pure dispersion), their molar cohesive energies and other properties can be computed reasonably accurately. Usually, however, two or more interactions occur simultaneously, and it becomes difficult to apply simple potential functions, especially when orientation-dependent dipolar and H-bonding interactions are involved. In spite of this complexity, some general patterns do emerge when we compare the boiling points of different compounds, which is a measure of the

TABLE XIII

Relative strengths of different types of interactions as reflected in the boiling points of different compounds[a]

Molecule		Molecular Weight	Dipole Moment (D)	Boiling Point (°C)
Ethane	CH_3CH_3	30	0	−89
Formaldehyde	HCHO	30	2.3	−21
Methanol	CH_3OH	32	1.7	64
n-Butane	$CH_3CH_2CH_2CH_3$	58	0	−0.5
Acetone	CH_3COCH_3	58	3.0	56.5
Acetic acid	CH_3COOH	60	1.5	118
n-Hexane	$CH_3(CH_2)_4CH_3$	86	0	69
Ethyl propyl ether	$C_5H_{12}O$	88	1.2	64
1-Pentanol	$C_5H_{11}OH$	88	1.7	137

[a] In order to make comparisons meaningful molecules have been put into three groups of similar molecular weights and size. Within each group the first molecule is nonpolar and interacts purely via dispersion forces, the second is polar, and the third interacts also via H-bonds.

cohesive forces holding molecules together in condensed phases (Chapter 2). Such a comparison is made in Table XIII, where we see that the weakest interactions are the dispersion and dipolar interactions, followed by H-bonding interactions (then there is a large jump to the much stronger covalent and ionic interactions, not shown in Table XIII). Note (i) the dominance of H-bonding forces (even when compared to very polar molecules such as acetone) and (ii) the increasing importance of dispersion forces for larger molecules. However, though dispersion forces are the ones mainly responsible for bringing molecules together, they lack the discrimination, specificity, and directionality of dipolar and H-bonding interactions, and it is these that ultimately determine the fine and subtle details of molecular and macro-molecular structures, such as those of molecular crystals, polypeptides (e.g., proteins), polynucleotides (e.g., DNA and RNA), micelles, and biological membranes.

8.5. The hydrophobic effect

The propensity of water molecules to form H bonds with each other has a strong influence on their interaction with nonpolar molecules that are incapable of forming H bonds (e.g., alkanes and other hydrocarbons). When water molecules come in contact with such a molecule they are faced with an apparent dilemma: whichever way the water molecules face, it would appear that one or more of the four charges per molecule will have to point towards the inert solute molecule and thus be lost to H-bond formation. Clearly the best configuration would have the least number of tetrahedral charges pointing towards the unaccommodating species so that the other charges can point towards the water phase and so be able to participate in H-bond attachments much as before. There are many options to salvaging lost H bonds. If the nonpolar solute molecule is not too large, it is possible for water molecules to pack around it without giving up *any* of their hydrogen-bonding sites. Examples of such arrangements are shown in Fig. 19. Since we have already established (Chapter 6) that the dispersion interaction between water

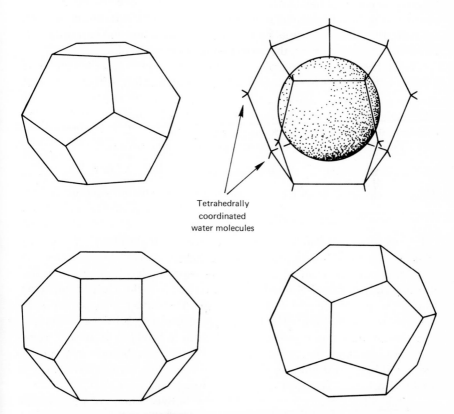

Tetrahedrally
coordinated
water molecules

Fig. 19 Clathrate "cages" formed by water molecules around a dissolved nonpolar solute molecule. Such structures are not rigid but labile, and their H bonds are not stronger than in pure water, but the water molecules forming these cages are more ordered than in the bulk liquid.

and hydrocarbons is not very different from that between water and water or, for that matter, between hydrocarbon and hydrocarbon, we see that the main effect of bringing water molecules and nonpolar molecules together is the *reorientation* of the water molecules so that they can participate in H-bond formation more or less as in bulk water (i.e., without necessitating any breakage of H bonds).

Thus, thanks to the uncanny ability of tetrahedrally coordinated molecules to link themselves together around almost any inert molecule, whatever its size or shape, the apparent dilemma mentioned earlier is often easily solved. However, the sizes and shapes of nonpolar solute molecules appear to be fairly critical in determining the water structure adopted around them; this is often referred to as *hydrophobic solvation* or *hydrophobic hydration*. At present there is no simple theory of such solute–solvent interactions. However, both theoretical and experimental studies do indicate that the reorientation, or restructuring, of water around nonpolar solutes or surfaces is *entropically* very unfavourable, since it disrupts the existing water structure and imposes a new, more ordered, structure on the surrounding water molecules.

It is for this reason that hydrocarbons are so sparingly soluble in water, with a highly unfavourable free energy of solubilization that is mainly entropic and, as we saw in Chapter 6, cannot be accounted for by continuum theories of van der Waals forces. For example, the free energy of transfer of methane and n-butane molecules from bulk liquid into water at $25°C$ is about 14.5 and 24.5 kJ mol^{-1}, respectively. For n-butane, this is split up as follows:

$$\Delta G = \Delta H - T \Delta S = -4.2 + 28.7 = +24.5 \quad kJ \, mol^{-1}.$$

Thus, the decrease in entropy contributes 85% to this interaction, and for many other hydrocarbons (e.g., benzene) the entropic contribution to ΔG is even higher—closer to 100%.

It is also observed that for different hydrocarbon molecules, the free energy of transfer is roughly proportional to the surface areas of the molecules, an indication that the number of reoriented water molecules is more or less determined by the non–H-bonding area exposed to them. Thus, for methane, whose van der Waals radius is $a = 0.2$ nm, the surface area per molecule is about $4\pi a^2 \approx 0.50$ nm^2. The free energy of transfer, $\Delta G = 14.5$ kJ mol^{-1}, when calculated per unit surface area turns out to be about 48 mJ m^{-2}. Similarly, for n-butane, the surface area is given by (see Section 7.1) $4\pi a^2 + 2\pi a(3 \times 0.127) \approx 1.0$ nm^2, so that the corresponding surface free energy is about 41 mJ m^{-2}. These values are very close to the interfacial free energies γ of liquid alkane–water interfaces. The high surface tension of water, about 72 mJ m^{-2}, may also be taken as an example of this mainly entropic effect since air behaves as a nonpolar, non–H-bonding medium in this sense. We shall return to considerations of surface energies and other surface and interfacial phenomena in later chapters.

The low solubility of nonpolar molecules in water and the mainly entropic nature of the free energy of solubilization is known as the *hydrophobic effect* (Kauzmann, 1959; Tanford, 1980), and substances such as hydrocarbons and fluorocarbons are known as *hydrophobic substances.** Similarly, *hydrophobic surfaces* are not "wetted" by water; instead, when water comes in contact with such surfaces it subtends a large contact angle (Chapter 14).

8.6. The hydrophobic interaction

Closely related to the hydrophobic *effect* is the hydrophobic *interaction*, which describes the unusually strong attraction between nonpolar (hydrophobic) molecules and surfaces in water—often stronger than their interaction in free space. For example, the van der Waals interaction between two contacting methane molecules in free space is about -2.5×10^{-21} J (from Table VIII), while in water the interaction energy is about -14×10^{-21} J. Similarly, the surface tensions of most hydrocarbons lie in the range 15–30 mJ m^{-2}, while their interfacial tensions with water are in the range 35–50 mJ m^{-2}. In Section 6.6 we saw that this strong interaction in water cannot be accounted for by continuum theories of van der Waals forces (which actually predict a *reduced* interaction in water). Because of its strength it was believed that some sort of "hydrophobic bond" was responsible for this interaction. But it should be clear from what has been described above that there is no bond associated with this mainly entropic phenomenon, which arises primarily from the structural rearrangement of water molecules in the overlapping solvation zones as two hydrophobic species come together (and is therefore of longer range than any typical bond).

To date there have been very few measurements of the hydrophobic interaction between dissolved nonpolar molecules, mainly because they are so insoluble. Tucker et al. (1981) reported values of -8.4 and -11.3 kJ mol^{-1} for the free energies of dimerization of benzene–benzene and cyclohexane–cyclohexanol, respectively, and Ben Naim et al. (1973) have deduced a value of about -8.5 kJ mol^{-1} for two methane molecules. On the theoretical side the problem is horrendously difficult, and there are no simple theories of the hydrophobic interaction though a number of promising theoretical approaches have been proposed (Dashevsky and Sarkisov, 1974; Pratt and

---

* Originally coined to describe the water-hating properties of these substances. It is well to note that their interaction with water is actually attractive, due to the dispersion force, though the interaction of water with itself is much more attractive. Water simply loves itself too much to let some substances get in its way.

Chandler, 1977; Marčelja *et al.*, 1977; Pangali *et al.*, 1979; reviewed by Nicholson and Parsonage, 1982).

Israelachvili and Pashley (1982b) measured the hydrophobic force law between two macroscopic curved hydrophobic surfaces in water and found that in the range 0–10 nm the force decayed exponentially with distance with a decay length of about 1 nm [an exponential distance-dependence for the interaction had previously been proposed by Marčelja *et al.* (1977)]. Based on these findings Israelachvili and Pashley proposed that for small solute molecules, the hydrophobic free energy of dimerization increases in proportion with their diameter σ according to

$$\Delta G \text{ (hydrophobic interaction)} \approx -20\sigma \quad kJ\, mol^{-1}, \qquad (8.1)$$

where σ is in nm. For example, for cyclohexane ($\sigma = 0.57$ nm), this gives $\Delta G \approx -11.4$ kJ mol^{-1} in agreement with the measured value of -11.3 kJ mol^{-1}.

The hydrophobic interaction plays a central role in many surface phenomena, in micelle formation, in biological membrane structure, and in determining the conformations of proteins (Kauzmann, 1959; Tanford, 1980; Israelachvili *et al.*, 1980a), and we shall encounter this interaction again in later chapters.

8.7. Hydrophilicity

While there is no phenomenon actually known as the hydrophilic effect or the hydrophilic interaction, such effects can be recognized in the propensity of certain molecules and groups to be water soluble and to strongly *repel* each other in water, in contrast to the strong *attraction* exhibited by hydrophobic groups. Hydrophilic (i.e., water loving) groups prefer to be in contact with water rather than with each other, and they are often *hygroscopic* (taking up water from vapour). As one might expect, strongly hydrated ions and zwitterions (discussed in Chapter 4) are hydrophilic; but some uncharged and even nonpolar molecules can be hydrophilic if they have the right geometry and if they contain electronegative atoms capable of associating with the H-bond network in water, for example, the O atoms in alcohols and polyethylene oxide and the N atoms in amines. Table XIV lists some common hydrophilic molecules and molecular groups, as well as some hydrophilic surfaces that are wetted by water or that repel each other in aqueous solutions. From this table we see that a polar group is not necessarily hydrophilic and that a nonpolar group is not always hydrophobic!

The degree of hydrophilicity of molecular groups is important for understanding the interactions and associations of *amphiphilic molecules* such as surfactants and lipids (Part III). In such molecules one end contains a

TABLE XIV

Hydrophilic groups and surfaces

Molecules and ions

Alcohols (CH_3OH, C_2H_5OH, glycerol)	Soluble proteins
Sugars (glucose, sucrose)	DNA
Polyethylene oxide ($-CH_2CH_2O-$)$_n$	Li^+, Mg^{2+}, Ca^{2+}, La^{3+}

Molecular groups[a]

Anionic		*Zwitterionic*	
Carboxylate	$-COO^-$	Phosphatidylcholine	$-\overset{+}{O}PO_2OCH_2CH_2\overset{+}{N}(CH_3)_3$
Sulfonate	$-SO_3^-$	(lecithin)	
Phosphate ester	$-OPO_2^-O-$		

Cationic		*Polar (nonionic)*	
Trimethyl ammonium	$-\overset{+}{N}(CH_3)_3$	Amine	$-NH_2$
		Amine oxide	$-NO(CH_3)_2$
		Sulfoxide	$-SOCH_3$
		Phosphine oxide	$-PO(CH_3)_2$

Polar groups that are not hydrophilic when attached to a long hydrocarbon (hydrophobic) chain[a]

Alcohol	$-OH$	Amide	$-CO \cdot NH_2$
Ether	$-OCH_3$	Nitrosoalkane	$-N{=}O$
Mercaptan	$-SH$	Aldehyde	$-CH{=}O$
Amines	$-NH(CH_3)$, $-N(CH_3)_2$	Ketone	$-CO \cdot CH_3$

Solid surfaces

Silica below 600°C (surface characterized by Si—OH groups)
Swelling clays (montmorillonite)
Gold (when clean)

[a] Complied from a longer list given by Laughlin (1978, 1981).

hydrophilic group while the rest of the molecule is hydrophobic (usually a long hydrocarbon chain). The hydrophilicity of some hydrophilic groups, for instance the OH group, can be completely neutralized when they are attached to a long n-alkyl chain such as $-(CH_2)_{11}CH_3$ (see Table XIV).

It appears therefore that the hydrophilic and hydrophobic interactions, unlike the dispersion interaction, are not additive. Indeed one would not expect them to be, since both are essentially solvation-type interactions that depend rather specifically on the structure of water adopted around dissolved groups. In Chapter 13 we shall see that both the hydrophobic and the hydrophilic (hydration) interactions as measured between macroscopic surfaces in water are of relatively long range, extending a few molecular diameters, and that when both hydrophilic and hydrophobic groups are present the net interaction is not the sum of the separate components.

PART TWO

THE FORCES BETWEEN PARTICLES
AND SURFACES

Chapter 9

Some Unifying Concepts in Intermolecular and Interparticle Forces

9.1. Factors favouring the association of like molecules or particles in a medium

In this second part we shall be looking at the physical forces between particles and surfaces. While the fundamental forces involved are the same as those already described (i.e., electrostatic, van der Waals, and solvation forces), they can manifest themselves in quite different ways and lead to qualitatively new features when acting between macroscopic particles or extended surfaces. These differences will be discussed in the next chapter. Here we shall look at the similarities, and see how the ideas developed in Part I also apply to the interactions of macroscopic particles and surfaces. We shall find that, independently of the specific type of interaction force involved, certain semiquantitative relations describing molecular forces—known as *combining relations*—are applicable quite generally to all systems (i.e., to the interactions of molecules, particles, surfaces, and even complex multicomponent systems).

Let us start by noting that for any type of interaction between two molecules A and B, the interaction energy at any given separation is always (to a good approximation) proportional to the product of some molecular property of A times some molecular property of B. Let us denote these properties by A and B. Referring to Fig. 4 we find, for example, that for the charge–nonpolar-molecule interaction, we may write $A \propto Q_A^2$ and $B \propto \alpha_B$; for the dipole–dipole interaction, $A \propto u_A$ or u_A^2 and $B \propto u_B$ or u_B^2; for the dispersion interaction, we have $A \propto \alpha_A$, $B \propto \alpha_B$. Note that even for the gravitational interaction [Eq. (1.1)], we may put $A \propto$ (mass of body A), $B \propto$ (mass of body B).

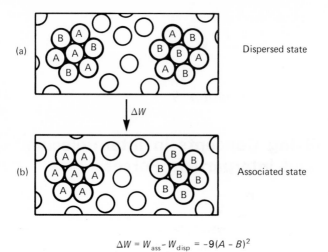

(a)   Dispersed state

ΔW

(b)   Associated state

$$\Delta W = W_{\text{ass}} - W_{\text{disp}} = -9(A - B)^2$$

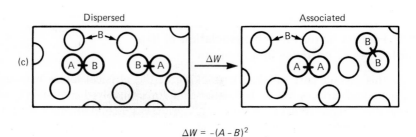

$$\Delta W = -(A - B)^2$$

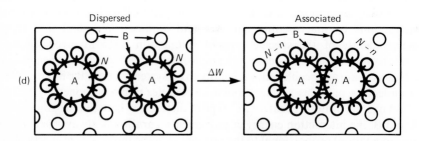

$$\Delta W = -n(A - B)^2$$

Fig. 20 (a) Two central molecules A and B surrounded by an equal number of A and B molecules in a dispersed solvent medium. Since there are three A–A "bonds," three B–B "bonds," and 18 A–B "bonds" we may write $W_{\text{disp}} = -(3A^2 + 3B^2 + 18AB)$.
(b) Seven A molecules and seven B molecules have associated in two small clusters. There are now 12 A–A "bonds" and 12 B–B "bonds" so that $W_{\text{ass}} = -12(A^2 + B^2)$. The net change in

Thus, for many different types of interactions, we may express the binding energies of molecules A and B in contact as

$$W_{AA} = -A^2, \qquad W_{BB} = -B^2 \qquad \text{(for like molecules)} \qquad (9.1)$$

and

$$W_{AB} = -AB \qquad \text{(for unlike molecules),} \qquad (9.2)$$

where only for the purely Coulombic charge–charge interaction are the negative signs reversed (cf. Fig. 4). Let us now consider a liquid consisting of a mixture of molecules A and B in equal amounts. If the molecules are randomly *dispersed* (Fig. 20a), then on average an A molecule will have both A and B molecules as nearest neighbours, and similary for molecule B. However, if the molecules are *associated*, then the molecular organization of nearest neighbours around molecules A and B will be as in Fig. 20b. The difference in energy between the associated and dispersed clusters will therefore be $\Delta W = -9(A - B)^2$, where we note that in this two-dimensional case nine A–A "bonds" and nine B–B "bonds" have replaced 18 A–B "bonds." In three dimensions with 12 nearest neighbours around each central molecule, and starting with six A and six B molecules around each A and B molecule, we would find that $\Delta W = -27(A - B)^2$ and that 27 A–A and 27 B–B "bonds" have been formed on association. For the simplest case of two associating A molecules (Fig. 20c), we have $\Delta W = -(A - B)^2$.

Thus in general we may write

$$\Delta W = W_{ass} - W_{disp} = -n(A - B)^2, \qquad (9.3)$$

where n is always equal to the number of like "bonds" that have been formed in the process of association, irrespective of how many molecules are involved or their relative sizes (Fig. 20d). Since $(A - B)^2$ must always be positive we see that in general $\Delta W < 0$ (i.e., $W_{ass} < W_{disp}$). We may therefore conclude that the associated state of like molecules is energetically preferred to the dispersed state. In other words, *there is always an effective attraction between like molecules or particles in a binary mixture.*

---

energy on going from the dispersed to the associated state is therefore $\Delta W = W_{ass} - W_{disp} = -9(A - B)^2$. Note that ΔW does not involve any term due to the interaction with the surrounding medium, which is assumed to remain unchanged during the redistribution of the A and B molecules.

(c) Two A molecules associating in a medium of B molecules. Here $W_{disp} = -2AB$, and $W_{ass} = -(A^2 + B^2)$, so that $\Delta W = -(A - B)^2$.

(d) Two large particles A associating in a medium of B molecules involving the replacement of $2n$ A–B "bonds" by n A–A "bonds" and n B–B "bonds". $W_{disp} = -2NAB$, $W_{ass} = -2(N - n)AB - nA^2 - nB^2$, so that $\Delta W = -n(A - B)^2$. Note that n is proportional to the effective area of adhesion of the two particles (roughly proportional to their radii).

Equation (9.3) can be developed further to provide additional insight into the interactions of like solute molecules and particles in a medium. First, Eq. (9.3) may be expressed in a number of different forms:

$$\Delta W = -n(A - B)^2 = -n(A^2 + B^2 - 2AB)$$
$$= -n(\sqrt{-W_{AA}} - \sqrt{-W_{BB}})^2 = n(W_{AA} + W_{BB} - 2W_{AB}), \tag{9.4}$$

where ΔW may be readily seen to be the same as the interaction pair potential $w(\sigma)$ in the medium (at contact). Second, since $-W_{AA}$ and $-W_{BB}$ are roughly proportional to the respective molar cohesion energies U_A and U_B we find that

$$\Delta W \propto -(\sqrt{U_A} - \sqrt{U_B})^2, \tag{9.5}$$

which is essentially the same as Eq. (6.39), previously derived for the specific case of dispersion forces. If W_{AA} and W_{BB} are sufficiently different (i.e., if the molecules are very different; for example, A polar, B nonpolar), then ΔW will be large enough to overcome the entropic drive to disorder resulting in a low solubility (immiscibility) or phase separation. The immiscibility of water and hydrocarbons and the "like-dissolves-like" rule (Sections 6.6 and 6.7) are examples of this phenomenon. Furthermore, since $\Delta W \propto n$, larger particles or polymers are more likely to phase separate than smaller particles, and indeed the vast majority of polymers are immiscible with each other. Third, the value of $\Delta W/n$ for molecules A coming into contact in medium B is the same as for the inverse case of molecules B associating in medium A. This reciprocity property was previously noted for the specific case of van der Waals forces (Section 6.6). Finally, Eq. (9.4) clearly shows that the interaction of two solute molecules in a solvent medium is intimately coupled to the strength of the solvent–solvent interaction. Thus, the attraction between two particles in water or between two protein molecules in a lipid bilayer is not independent of the surrounding water–water or lipid–lipid interactions.

While the above semiquantitative criteria have broad applicability, there are two very important exceptions: First, for the Coulomb interaction between charged atoms or ions, since the sign of ΔW is reversed, the dispersed state (Fig. 20a) is the favoured one. Thus, in ionic crystals (e.g., Na^+Cl^-), the cations and anions are nearest neighbours in the lattice. Second, hydrogen-bonding molecules do not readily fall into this simple behaviour pattern because the strength of the H bond between two different molecules cannot be expressed simply in terms of $W_{AB} = -AB$. For example, while an acetone molecule cannot form H bonds with another acetone molecule, it can do so with water via its $C{=}O$ group, and for this reason, acetone is miscible with water. Such strongly hydrophilic groups or molecules (see Table XIV) will repel each other in water due to their strong binding to water, and their specific solvation interactions in water (Chapter 13) cannot be described in terms of the simple equations presented here.

9.2. Two like surfaces coming together in a medium

The above relations may be readily applied to the interaction of two extended surfaces in a liquid medium. For two associating surfaces A in medium B, the number n of bonds involved is simply proportional to their surface area of interaction (cf. Fig. 20d). W_{AA} is now proportional to the so-called *cohesion energy* $2\gamma_A$ of two unit areas of A coming into contact in vacuum, and similarly for W_{BB}. W_{AB} is proportional to the *adhesion energy* of two unit areas of A and B coming together, while ΔW is proportional to the *interfacial energy* $2\gamma_{AB}$ per unit area of the A–B interface. These surface energies will be discussed in more detail in Chapter 14; for the moment, let us note that for two surfaces, Eq. (9.4) may now be expressed in the form

$$\gamma_{AB} = \gamma_A + \gamma_B - W_{AB} \qquad \text{per unit area.} \qquad (9.6)$$

This important thermodynamic relation is valid for both solid and liquid interfaces. It gives the free energy (always negative) of bringing unit areas of surfaces A into contact in liquid B, and vice versa, since $\gamma_{AB} = \gamma_{BA}$.

All the above equations belong to an important class of expressions known as *combining relations* or *combining laws*. They are extremely useful for deriving relationships between various energy terms in a complex system and are often used for obtaining approximate values for parameters that cannot be easily measured. For example, if we return to Eq. (9.4), we may also write it as

$$\Delta W = W_{AA} + W_{BB} - 2\sqrt{W_{AA}W_{BB}} \qquad (9.7)$$

so that Eq. (9.6) becomes

$$\gamma_{AB} = \gamma_A + \gamma_B - 2\sqrt{\gamma_A\gamma_B} = (\sqrt{\gamma_A} - \sqrt{\gamma_B})^2, \qquad (9.8)$$

a useful expression that is often used to estimate the interfacial energy γ_{AB} solely from the surface energies or surface tensions of the pure liquids, γ_A and γ_B, in the absence of any data on the energy of adhesion W_{AB}. We shall encounter these and other combining relations again.

9.3. Factors favouring the association of unlike molecules, particles, or surfaces in a third medium

Let us now proceed with mixtures containing three different species (Fig. 21). For two unlike molecules or particles A and B coming together in a medium composed of molecules of type C (Fig. 21a,b), we find

$$\Delta W = W_{ass} - W_{disp} \propto -AB - C^2 + AC + BC$$

$$\propto -(A - C)(B - C). \qquad (9.9)$$

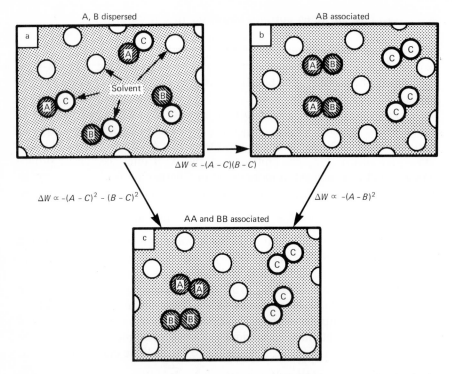

Fig. 21 (a) and (b)   Two unlike molecules or particles A and B may attract or repel each other in third medium C. Repulsion will occur (ΔW positive) if the properties of C are intermediate between those of A and B; for example, for gravitational forces, if the density of C (e.g., water) is between that of A (e.g., iron) and B (e.g., wood). In such cases the dispersed state (a) is energetically favoured.

(c)   The associated state of like molecules has a lower energy than either (a) or (b).

This is a very interesting result because it shows that the energy of association can now be either positive or negative. If positive, the particles will remain dispersed and so will effectively repel each other in the solvent medium C.

Let us see how an effective repulsion has arisen from interaction potentials that are all attractive to begin with. The phenomenon may be referred to as *Archimedes' principle* as applied to intermolecular forces. In gravitational forces we all know that iron sinks while wood rises in water. Thus, wood is effectively experiencing a repulsion from the earth when in water. This is because it is lighter than water, and if it were to descend, it would have to displace an equal volume of water and drive it upwards to replace the space previously occupied by the wood. Since water is denser and so more attracted to the earth than wood the net effect would be unfavourable, viz., the energy lost in water going up is not recovered by the energy gained in wood coming down. This displacement principle applies to all interactions including intermolecular interactions.

Equation (9.9) tells us that if C is intermediate between A and B, two particles (or surfaces) will repel each other, an effect that was previously noted for van der Waals forces (Section 6.6) where the operative properties of A, B, and C are their dielectric constants and refractive indices. However, whatever the relation between A, B, and C, the most favoured final state will be that of particles A associating with particles A, B with B, and C with C (Fig. 21c).

This procedure may be extended to mixtures with more species. We may therefore generalize our earlier conclusion, viz., *there is always an effective attraction between like molecules or particles in a multicomponent mixture* (again with the proviso that strong Coulombic or H-bonding forces are not involved). And in addition, *unlike particles may attract or repel each other in a solvent.*

9.4. Particle–surface interactions

Let us now extend the above analysis to the case of a particle C near an interface dividing two immiscible liquid media A and B (Fig. 22). Three situations may arise:

(i) the particle may be repelled from the interface on either side of it (desorption or negative adsorption),
(ii) it may be attracted to the interface from either side and remain there (adsorption), or
(iii) it may be attracted from one side but repelled from the other (engulfing).

Now using Eq. (9.9) we may write for the energy change for the particle coming up to the interface:

$$\text{from the left:} \quad \Delta \vec{W} \propto -(C - A)(B - A); \quad (9.10)$$

$$\text{from the right:} \quad \Delta \overleftarrow{W} \propto -(C - B)(A - B). \quad (9.11)$$

First, we note that

$$\Delta \overleftarrow{W}_{\text{tot}} \propto \Delta \overleftarrow{W} - \Delta \vec{W} \propto (A - C)^2 - (B - C)^2 \propto \gamma_{\text{BC}} - \gamma_{\text{AC}}, \quad (9.12)$$

which gives the net free energy change on taking the particle *across* the interface (Fig. 22b). Second, if $A > C > B$ or $A < C < B$, the particle will be attracted to the interface from either side, leading to *adsorption*. As can be seen,

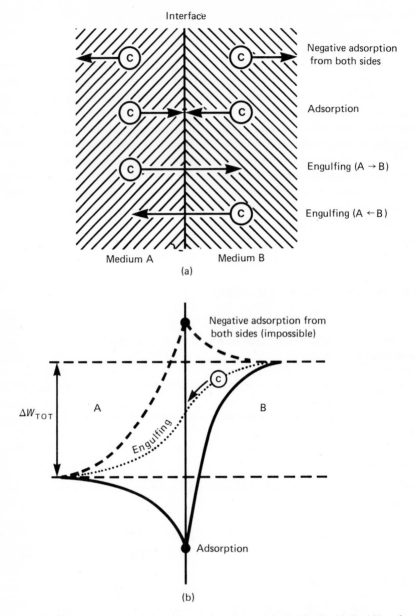

Fig. 22 (a) The three possible modes of interaction of a particle C with a liquid–liquid interface. (b) Corresponding schematic energy versus distance profiles (assumed monotonic) for $\Delta \overline{W}_{TOT} < 0$. Note that if medium A is solid, then particle C will adsorb on it from B since it cannot be engulfed by A.

this will occur if the particle's properties are intermediate between those of the two solvent media. The adsorption of amphiphilic molecules at hydrocarbon–water interfaces is an example of such a phenomenon. Amphiphilic molecules such as surfactants are partly hydrophilic and partly hydrophobic and so have properties intermediate between the two liquids. Third, if $A > B > C$ or $A < B < C$, the particle will be attracted to the interface from the left but repelled from the right, while if $B > A > C$ or $B < A < C$, it will be attracted from the right but repelled from the left (*engulfing*). Now since these six combinations exhaust all the possibilities, we see that repulsion from both sides of an interface (i.e., negative adsorption from both sides) cannot occur and that either adsorption or engulfing will be the rule. It is for this reason that surfaces are so prone to adsorbing molecules or particles from vapour or solution. Indeed, if medium B is a vapour ($B \approx 0$), then we must have $A > C > B$ or $C > A > B$, both of which result in a net attraction of gaseous molecules C towards the surface of A (i.e., vapour molecules will always be attracted to a surface).

9.5. Adsorbed surface films

The above examples and Fig. 22 apply only to isolated molecules of C; in other words, they apply only to dilute concentrations of C in media A and B and to monolayer or submonolayer adsorption. At higher concentrations the molecules or particles of C will associate into a separate phase to one or the other side of the A–B interface, or build up at the interface, depending on the relative magnitudes of A, B, and C. In this final section we shall consider the factors favouring the formation of a thick adsorbed film on a solid surface.

In Fig. 23a we have, initially, a solid surface of A in contact with a liquid mixture of molecules C in B. Again there are three possibilities:

(i) For $A > C > B$ or $A < C < B$ (C intermediate between A and B), molecules C will be attracted to A while molecules B will be repelled from it, and an adsorbed monolayer or film of C will be energetically favourable (Fig. 23b).

(ii) For $C > B > A$ or $C < B < A$ (B intermediate between A and C), the roles of B and C are reversed, favouring adsorption of B or negative adsorption of C (Fig. 23c).

(iii) Finally, when $B > A > C$ or $B < A < C$ (A intermediate), both molecules B and C are attracted to the solid surface. Under such circumstances no uniformly adsorbed film will form but different regions of the interface will collect macroscopic droplets of the A or B phase (Fig. 23d).

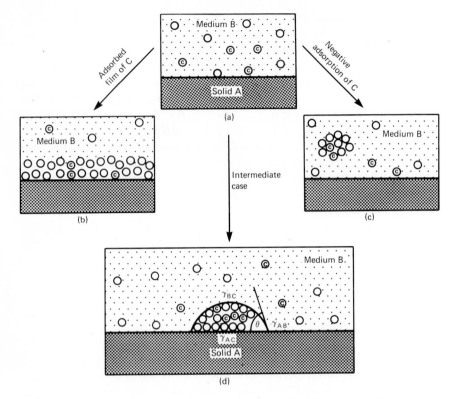

Fig. 23 (a) Low concentration of solute molecules C in medium B (i.e., below saturation).
(b) Near saturation, an adsorbed film of C develops and builds up in thickness as the concentration of C in B approaches saturation. [Corresponds to $\cos \theta > 1$ in Eq. (9.13).]
(c) Negatively adsorbed molecules of C on A (i.e., repulsion between C and A in medium B). [Corresponds to $\cos \theta < -1$ in Eq. (9.13).]
(d) Intermediate case, corresponds to $1 > \cos \theta > -1$.

It is left as an exercise for the interested reader to ascertain that when the energy of the whole system is minimized, the *contact angle* θ formed by these droplets (Fig. 23d) is given by

$$\cos \theta = (B + C - 2A)/(B - C), \qquad (9.13a)$$

which leads to values of $\cos \theta$ between 1 and -1, or θ between $0°$ and $180°$, only for A intermediate between B and C. The above may also be written in the forms

$$\gamma_{AC} + \gamma_{BC} \cos \theta = \gamma_{AB} \qquad \text{(Young equation)}, \qquad (9.13b)$$

$$\gamma_{BC}(1 + \cos \theta) = \Delta W \qquad \text{(Young–Dupré equation)}, \qquad (9.13c)$$

where ΔW is the adhesion energy per unit areas of surfaces A and C adhering in medium B. These important fundamental equations will be derived and discussed further in Chapter 14.

The purpose of the phenomenological discussion of this chapter is to illustrate how a few basic notions concerning two-particle interaction energies can be applied to progressively more complex situations, and vice versa [i.e., how fairly complex situations can arise from, and be understood in terms of, the simplest possible pair potential, Eq. (9.2)]. However, this type of nonspecific approach, while conceptually useful, has its limits in that it does not take into account the way interaction energies vary with distance. Two particles or surfaces may have an energy minimum at contact, but if the force law is not monotonic—it may be repulsive before it becomes attractive closer in—the particles will remain separated (i.e., they effectively repel each other). Only a quantitative analysis in terms of the operative forces and their distance dependence can provide a full understanding of interparticle and interfacial phenomena.

Chapter 10

Contrasts between Intermolecular, Interparticle, and Intersurface Forces

10.1. Short-range and long-range forces

We are told that when an apple fell on Newton's head it set in motion a thought process that eventually lead Newton to formulate the gravitational force law. The conceptual breakthrough in this discovery was the recognition that the force that causes apples to fall is the same force that holds the moon in its orbit round the earth.

On earth, gravity manifests itself in many different ways: in determining the height of the atmosphere, the capillary rise of liquids, and the behaviour of waves and ripples; in biology it decrees that animals that live in the sea (where the effect of gravity is almost negligible) can be larger than the largest possible land animal, that heavy land animals such as elephants must have short thick legs while a man or a spider can have proportionately long thin legs, that larger birds must have progressively larger wings (e.g., eagles and storks) while smaller birds, flies, and bees can have relatively small light wings, that only small animals can carry many times their own weight (e.g., ants), and much else (Thompson, 1968). But beyond the immediate vicinity of the earth's surface and out to the outer reaches of space this same force now governs the orbits of planets, the shapes of nebulae and galaxies, the rate of expansion of the universe, and, ultimately, its age. The first group of phenomena—those occurring locally on the earth's surface—may be thought of as the *short-range* effects of gravitational forces, while the second and very different types of phenomena occurring on a cosmological scale may be thought of as the *long-range* effects of gravity.

Intermolecular forces are no less versatile in the way in which the same force can have very different effects at short and long range, though here *short range* usually means at or very close to molecular contact (< 1 nm) while *long-range* forces are rarely important beyond 100 nm. In Part I we saw that the properties of gases and the cohesive strengths of condensed phases are determined mainly by the interaction energies of molecules in contact $\omega(\sigma)$ (i.e., molecules interacting with their immediate neighbours). For example, the van der Waals pair energy of two neighbouring molecules is at least 64 times stronger than that between next nearest neighbours [$1/\sigma^6$ compared to $1/(2\sigma)^6$]. Only the Coulomb interaction is effectively long range in that the energy decays slowly, as $1/r$, and remains strong at large distances. However, in a medium of high dielectric constant such as water we have seen how the strength of the Coulomb interaction can be much reduced.

We may therefore conclude that (apart from ionic crystals; see Chapter 3) the properties of a condensed phase are determined mainly by the strength of the interaction at or very near molecular contact, the long-range nature of the interaction (e.g., the exact distance dependence of the force law) playing only a minor role.

A very different situation arises when we consider the interactions of large particles or surfaces, for now when all the pair potentials between the molecules in each body is summed we shall find (i) that the net interaction energy is proportional to the radius of the particle and (ii) that it is very large at contact (many kT), and still appreciable at large separations up to 100 nm or more. In particular, when the force law is not monotonic (i.e., purely attractive or repulsive), then all manner of behaviour may arise depending on the specific form of the long-range distance dependence of the interaction, as illustrated in Fig. 24. For example, consider the purely attractive energy law in Fig. 24a. If both molecules and particles experience the same type of interaction, then both will be attracted to each other, and the thermodynamic properties of an assembly of molecules (in the gas or condensed phase) will be determined by the depth of the potential well at contact, as will the adhesion energy of two particles. However, for the type of energy law in Fig. 24b, two molecules will still attract each other since the energy barrier is negligibly small compared to kT; but two particles will effectively repel each other since the energy barrier is now too high to surmount. Under such circumstances particles dissolved in a medium will remain separated even though the ultimate *thermodynamic equilibrium state* is the aggregated state. Figure 24 illustrates some other types of commonly occurring intermolecular and intersurface potential functions and the different effects they have on molecule–molecule and particle–particle

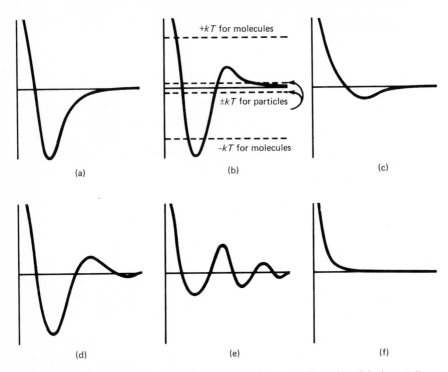

Fig. 24 Typical interaction potentials encountered between molecules and particles in a medium.
(a)   This potential is typical of vacuum interactions but is also common in liquids. Both molecules and particles attract each other.
(b)   Molecules attract each other; particles repel each other.
(c)   Weak minimum. Molecules repel, particles attract.
(d)   Molecules attract strongly, particles attract weakly.
(e)   Molecules attract weakly, particles attract strongly.
(f)   Molecules repel, particles repel.

interactions. In the following chapters we shall see how such interaction potentials can arise in different systems.

The interplay of the effects arising from the short-range and long-range part of an interaction can be very subtle. For example, the size and shape of small molecular aggregates such as surfactant micelles in water depend on the short-range interactions between the molecules, which are sensitive to electrolyte concentration, pH, temperature, etc. But so is the long-range interaction between these aggregates. The interdependence of short-range and long-range forces and, consequently, of intraparticle and interparticle forces is particularly important for "soft" structures, such as surfactant micelles, lipid vesicles, and biological membranes, discussed in Part III.

10.2. Interaction potentials between large bodies

In this section we shall investigate (semiquantitatively) how the strength of the pair interaction between two molecules compares with those occurring between a molecule and a surface, between a spherical particle and a surface, and between two flat surfaces.

Molecule–surface interaction

Let us once again assume that the pair potential between two atoms or small molecules is purely attractive and of the form $w(r) = -C/r^n$. Then, with the further assumption of *additivity*, the net interaction energy of a molecule and the planar surface of a solid made up of like molecules (Fig. 25a) will be the sum of its interactions with all the molecules in the body. For molecules in a circular ring of cross-sectional area $dx\,dz$ and radius x, the ring volume is $2\pi x\,dx\,dz$, and the number of molecules in the ring will be $2\pi\rho x\,dx\,dz$, where ρ is the number density of molecules in the solid. The net interaction energy for a molecule at a distance D away from the surface will therefore be

$$w(D) = -2\pi C\rho \int_{z=D}^{z=\infty} dz \int_{x=0}^{x=\infty} \frac{x\,dx}{(z^2 + x^2)^{n/2}} = -\frac{2\pi C\rho}{(n-2)} \int_D^\infty \frac{dz}{z^{n-2}}$$

$$= -2\pi C\rho/(n-2)(n-3)D^{n-3} \quad \text{for} \quad n > 3, \tag{10.1}$$

which for $n = 6$ (van der Waals forces) becomes

$$w(D) = -\pi C\rho/6D^3. \tag{10.2}$$

The corresponding force, $F = -\partial w(D)/\partial D = -\pi C\rho/2D^4$, could of course have been derived in a similar way by summing (integrating) all the pair forces resolved along the z axis.

Sphere–surface and sphere–sphere interaction

We can now calculate the interaction energy of a large sphere of radius R and a flat surface (Fig. 25b). First, from elementary geometry we know that for a circle

$$x^2 = (2R - z)z. \tag{10.3}$$

The volume of a thin circular section of area πx^2 and thickness dz is therefore $\pi x^2\,dz = \pi(2R - z)z\,dz$, so that the number of molecules contained within this section is $\pi\rho(2R - z)z\,dz$, where ρ is the number density of molecules in the sphere. Since all these molecules are at a distance $(D + z)$ from the planar

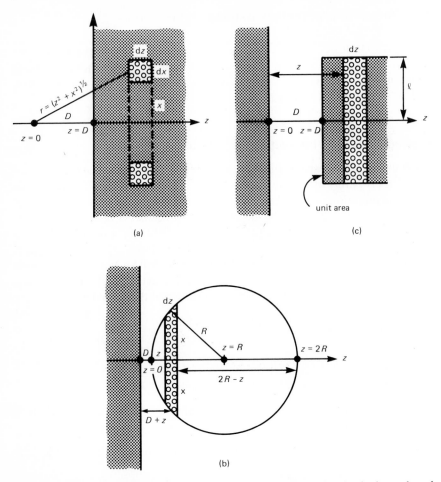

Fig. 25 Methods of summing (integrating) the interaction energies of molecules in condensed phases to obtain the interaction energies between macroscopic bodies.
(a)   Molecule near a wall.
(b)   A spherical particle near a wall ($R \gg D$).
(c)   Two planar surfaces ($l \gg D$).

surface, the net interaction energy is, using Eq. (10.1),

$$W(D) = -\frac{2\pi^2 C\rho^2}{(n-2)(n-3)} \int_{z=0}^{z=2R} \frac{(2R-z)z\,dz}{(D+z)^{n-3}}. \tag{10.4}$$

For $R \gg D$, only small values of z ($z \approx D$) contribute to the integral, and we obtain*

$$W(D) = -\frac{2\pi^2 C\rho^2}{(n-2)(n-3)} \int_0^\infty \frac{2Rz\,dz}{(D+z)^{n-3}}$$

$$= -\frac{4\pi^2 C\rho^2 R}{(n-2)(n-3)(n-4)(n-5)D^{n-5}}, \qquad (10.5)$$

which for $n = 6$ (van der Waals forces) becomes

$$W(D) = -\pi^2 C\rho^2 R/6D. \qquad (10.6)$$

However, for $D \gg R$, we may replace $(D + z)$ in the denominator of Eq. (10.4) by D, and we then obtain

$$W(D) = -\frac{2\pi^2 C\rho^2}{(n-2)(n-3)} \int_0^{2R} \frac{(2R-z)z\,dz}{D^{n-3}} = -\frac{2\pi C\rho(4\pi R^3\rho/3)}{(n-2)(n-3)D^{n-3}},$$

which, since $4\pi R^3\rho/3$ is simply the number of molecules in the sphere, is the same as Eq. (10.1) for the interaction of a molecule (or small sphere) with a surface. It is left as an exercise for the interested reader to show that for two spheres of equal radii R whose surfaces are at a small distance D apart $(R \gg D)$, their interaction energy is one half that given by Eq. (10.5) or (10.6), while for two spheres far apart $(D \gg R)$ the energy varies as $-1/D^n$ as for two molecules. At intermediate separations $(R \approx D)$ the expression for the interaction potential is more complicated but remains analytic (Hamaker, 1937).

Surface–surface interactions

Let us now calculate the interaction energy of two planar surfaces a distance D apart. For two infinite surfaces, the result will be infinity, and so we have to consider the energy per unit surface area. Let us start with a thin sheet of molecules of unit area and thickness dz at a distance z away from an extended surface of *larger* area (Fig. 25c). From Eq. (10.1) the interaction energy of this sheet with the surface is $-2\pi C\rho(\rho\,dz)/(n-2)(n-3)z^{n-3}$. Thus, for the two surfaces, we have

$$W(D) = -\frac{2\pi C\rho^2}{(n-2)(n-3)} \int_D^\infty \frac{dz}{z^{n-3}} = -\frac{2\pi C\rho^2}{(n-2)(n-3)(n-4)D^{n-4}}, \qquad (10.7)$$

---

* To avoid confusion we shall use W and D to denote the interaction free energies of macroscopic bodies whose surfaces are at a distance D apart, reserving w and r for the interactions of molecules.

which for $n = 6$ becomes

$$W(D) = -\pi C \rho^2/12D^2 \quad \text{per unit area.} \tag{10.8}$$

It is important to note that Eqs. (10.7) and (10.8) are for unit area of one surface interacting with an infinite area of another surface. In practice this usually amounts to two unit areas of both surfaces, but it is strictly applicable only when D is small compared to the lateral dimensions of the surfaces.

When two large spheres or a sphere and a wall are close together, one sometimes wants to know what is their "effective area" of interaction. First, we may note that no matter how large a sphere becomes it *never* approaches the behaviour of a flat surface. Its interaction energy will increase linearly with radius R [Eq. (10.5)], but the distance dependence of the interaction will not change (see also Fig. 27). However, the concept of an effective "interaction area" or "interaction zone" is a useful one. If we compare Eq. (10.5) for a sphere near a surface with Eq. (10.7) for two surfaces, we find that at the same surface separation D the effective area A for the sphere–surface interaction is

$$A = 2\pi RD/(n - 5)$$

$$= 2\pi RD \quad \text{for} \quad n = 6.$$

Now from Fig. 25b and Eq. (10.3) this area is simply equal to πx^2 when $z = D$ (so long as $R \gg D$). In other words, the effective area of interaction of a sphere with a surface or with another sphere is the circular zone whose centre is at a distance D from the surface (inside the sphere). The effective area decreases linearly with D as may be expected. For example, for a sphere of radius $R = 1$ μm at a distance $D = 1$ nm from a wall, the effective interaction area is about $2\pi RD \approx 10^{-14}$ m^2 or 10^4 nm^2, which corresponds to an area of radius ~ 45 nm.

The geometries considered above are the most commonly encountered. In the following chapter we shall look specifically at the van der Waals interactions of these and of other geometries as well (e.g., cylinders). Meanwhile, let us scrutinize some of the implications of the interaction potentials obtained so far.

10.3. Interactions of large bodies compared to those between molecules

First we may note that for two macroscopic bodies, the interaction energy decays much more slowly with distance than it does for two molecules. For example, whereas the van der Waals energy between atoms and molecules is of short range, having an inverse sixth-power distance dependence, the van der

Waals energy between large condensed bodies decays much more slowly with distance (cf. $1/D$ for spheres, $1/D^2$ for planar surfaces) and is effectively of much longer range. This is yet another manifestation of the long-range nature of interparticle forces.

Second, the van der Waals interaction energy of a small molecule of diameter σ with a wall is given by Eq. (10.2) as $w(D) = -\pi C\rho/6D^3$. Thus, at contact we may put $D \approx \sigma$, and putting $\rho \approx \sqrt{2}/\sigma^3$ (corresponding to a close packed solid), we obtain

$$w(\sigma) \simeq -\sqrt{2}\pi C/6\sigma^6 \simeq -0.74C/\sigma^6, \qquad (10.9)$$

which is of the same order as $-C/\sigma^6$ for two small molecules in contact! Likewise, for a sphere of atomic dimensions ($R = \sigma/2$) in contact with a wall ($D = \sigma$), we find from Eq. (10.6) that

$$W(\sigma) \approx -2\pi^2 C/12\sigma^6 \approx -1.6C/\sigma^6, \qquad (10.10)$$

while for two spheres

$$W(\sigma) \approx -0.8C/\sigma^6, \qquad (10.11)$$

which is very close to the molecule–molecule pair potential at contact. However, once the size of a sphere increases above atomic dimensions (i.e., once $R > \sigma$), then at contact ($D = \sigma$) Eq. (10.6) becomes

$$W(\sigma) \approx -2\pi^2 CR/6\sigma^7 \approx -1.6(2R/\sigma)C/\sigma^6, \qquad (10.12)$$

which reduces to Eq. (10.10) only for small radii, of order $R \approx \sigma/2 \approx 0.1–0.2$ nm, but which increases linearly with R for larger spheres; likewise for the interaction of two spheres. This is an important result. It shows that for diameters beyond about 0.5 nm, a molecule must already be considered as a (small) particle or else the strength of its interaction will be underestimated. The higher strength of the measured interaction between large molecules over that predicted on the assumption that $D = 2R$ was seen in Table VIII for CCl_4 but not for the smaller molecules. In conclusion, while for two atoms or small molecules the contact interaction energy has no explicit size dependence, that between larger particles increases linearly with their radius.

As regards the stabilizing repulsive forces between large bodies, these too manifest themselves in quite different ways when acting between macroscopic bodies. While the short-range steric or Born repulsion between individual molecules remains the same, large spheres will deform elastically when in contact under a large adhesive force. In other words, they will not remain spherical unless the attraction is weak or the material is very rigid. The interactions of deformable curved surfaces (e.g., elastic spheres or bilayer vesicles) will be discussed in Chapters 14 and 17.

10.4. Interaction energies and interaction forces: the Derjaguin approximation

So far we have been dealing mainly with interaction *energies* rather than the *forces* experienced by molecules and particles. This is because most experimental data on molecular interactions are readily understood in terms of interaction energies, as we saw in Part I. However, the forces between macroscopic bodies are often of greater interest, and are easier to measure, than their interaction energies. The two geometries most commonly encountered in practice are two flat surfaces or two spheres, a sphere near a flat surface being a special case of two spheres with one sphere very much larger than the other. On the theoretical side it is usually easier to derive expressions for the interaction potentials of two planar surfaces rather than for curved surfaces. Consequently it is desirable to be able to relate the force law $F(D)$ between two curved surfaces (e.g., spheres) to the interaction free energy $W(D)$ of two planar surfaces. A glance at Eq. (10.5) shows that for the additive intermolecular pair potential $w(r) = -C/r^n$, the value of $F(D)$ for a sphere near a flat surface is

$$F(D) = -\frac{\partial W(D)}{\partial D} = -\frac{4\pi^2 C\rho^2 R}{(n-2)(n-3)(n-4)D^{n-4}}. \tag{10.13}$$

This force law can be seen to be simply related to $W(D)$ per unit area of two planar surfaces, Eq. (10.7), by

$$F(D)_{\text{sphere}} = 2\pi R W(D)_{\text{plane}}. \tag{10.14}$$

This is a very useful relationship, and while it was derived for the special case of an additive inverse power potential, it is in fact valid for any type of force law, as will now be shown.

Assume that we have two large spheres of radii R_1 and R_2 a small distance D apart (Fig. 26). If $R_1 \gg D$ and $R_2 \gg D$, then the force between the two spheres can be obtained by integrating the force between small circular regions of area $2\pi x\,dx$ on one surface and the opposite surface, which is assumed to be a locally *flat* surface at a distance $Z = D + z_1 + z_2$ away. The net force between the two spheres (in the z direction) is therefore

$$F(D) \simeq \int_{Z=D}^{Z=\infty} 2\pi x\,dx\,f(Z), \tag{10.15}$$

where $f(Z)$ is the normal force per unit area between two flat surfaces. Since $x^2 \approx 2R_1 z_1 = 2R_2 z_2$, we have

$$Z = D + z_1 + z_2 = D + \frac{x^2}{2}\left(\frac{1}{R_1} + \frac{1}{R_2}\right)$$

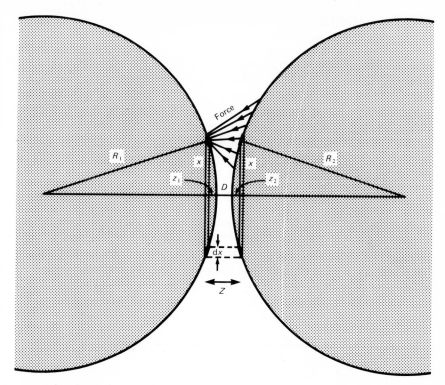

Fig. 26 The Derjaguin approximation (Derjaguin, 1934), which relates the force law $F(D)$ between two spheres to the energy $W(D)$ per unit area of two flat surfaces by $F(D) = 2\pi[R_1R_2/(R_1 + R_2)]W(D)$.

and

$$dZ = \left(\frac{1}{R_1} + \frac{1}{R_2}\right)x\,dx,$$

so that Eq. (10.15) becomes

$$F(D) \approx \int_D^\infty 2\pi\left(\frac{R_1R_2}{R_1 + R_2}\right)f(Z)\,dZ = 2\pi\left(\frac{R_1R_2}{R_1 + R_2}\right)W(D), \quad (10.16)$$

which gives the force between two spheres in terms of the energy per unit area of two flat surfaces at the same separation D. Eq. (10.16) is known as the *Derjaguin approximation*. It is applicable to any type of force law, whether attractive, repulsive, or oscillatory, so long as the range of the interaction and the separation D is much less than the radii of the spheres. It has been well verified experimentally, as discussed in Chapters 11–14. From Eq. (10.16) we may note the following:

(i) If one sphere is very large so that $R_2 \gg R_1$, we obtain $F(D) = 2\pi R_1 \cdot W(D)$, which is the same as Eq. (10.14) and corresponds to the limiting case of a sphere near a flat surface.

(ii) For two equal spheres of radii $R = R_1 = R_2$, we obtain $F(D) = \pi R W(D)$, which is half the value for one sphere near a flat surface.

(iii) For two spheres in contact $(D = \sigma)$, the value of $W(\sigma)$ can be associated with 2γ, where γ is the conventional *surface energy* per unit area of a surface. Eq. (10.16) then becomes

$$F(\sigma) = F_{ad} = 4\pi\{R_1 R_2/(R_1 + R_2)\}\gamma, \tag{10.17}$$

which gives the adhesion force F_{ad} between two spheres in terms of the surface energies of the materials. Adhesion forces are discussed in Chapter 14.

(iv) Perhaps the most intriguing aspect of the Derjaguin approximation is that it tells us that the distance dependence of the force between two curved surfaces can be quite different from that when the same type of force is acting between two flat surfaces. This is illustrated in Fig. 27, where we see that a purely repulsive force between curved surfaces can be attractive between two planar surfaces (over a certain distance regime) with equilibrium at some finite separation (Fig. 27b). Conversely, a purely attractive force can become repulsive between two planar surfaces (Fig. 27c).

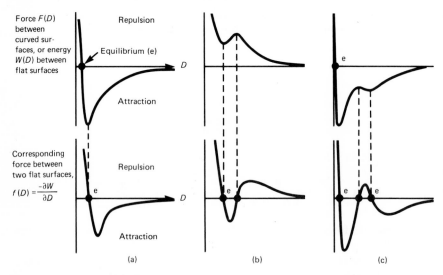

Fig. 27 Top row: force laws between two curved surfaces (e.g., spherical particles). Bottom row: corresponding force laws between two flat surfaces.

(v)   Finally, it may be readily shown that for two cylinders of radii R_1 and R_2 crossed at right angles to each other, the Derjaguin approximation becomes

$$F(D) = 2\pi\sqrt{R_1 R_2}\, W(D). \tag{10.18}$$

For two equal cylinders of radii $R = R_1 = R_2$, the above reduces to the same result as for a sphere of radius R near a flat surface, Eq. (10.14). In other words the interaction of two crossed cylinders is the same as that of a sphere and a wall if all three radii are the same.

10.5. Experimental measurements of intermolecular and surface forces

When we come to consider experimental measurements of intermolecular forces we are confronted with a bewildering variety of data to draw upon. This is because almost any measurement in the below-10-eV category, whether in physics, chemistry, or biology, is in some respect a measurement of intermolecular forces. For example, we have already seen how such a common property as the boiling point of a substance provides information on the strength of intermolecular contact energies. It is therefore difficult to make a list of experimental measurements of intermolecular forces; it is far better to draw upon whatever relevant data exists as the situation arises. This was done in Part I, and we shall continue with this practice in Parts II and III. However, different types of measurements do provide different insights and information, and in the rest of this section we shall categorize experiments according to the type of information they provide on intermolecular and intersurface interactions. First, let us recapitulate some of those already mentioned.

(i)   Thermodynamic data on gases, liquids, and solids (e.g., PVT data, boiling points, latent heats of vapourization, lattice energies) provide information on the short-range attractive energies between molecules. Measurements of adsorption isotherms and the surface potentials of solid–liquid interfaces provide information on the (adsorption) energies of molecules and ions interacting with surfaces.

(ii)   Physical data on gases, liquids, and solids (e.g., molecular beam scattering experiments, viscosity, compressibility, x-ray and neutron scattering of liquids and solids) provide information on the short-range interactions of molecules, especially their repulsive forces (which give insight into molecular size and shape), and the structure of condensed phases.

(iii)   Thermodynamic data on liquid mixtures (e.g., solubility, partitioning, miscibility, osmotic pressure) provide information on short-range solute–solvent and solute–solute interactions.

Such experimental data as listed in (i)–(iii) above often provide thermodynamic information only, so that direct access to the intermolecular potential functions (i.e., their distance dependence) is not possible. Thus, experimental *PVT* data may be compared with the van der Waals equation of state (Chapter 6), which contains terms to account for both the attractive and repulsive forces, but it does not give any information on the nature and range of the force laws themselves. To gain this information, some direct measurement of forces is required. Of the various methods that have been devised for measuring molecular forces, the most direct employ macroscopic bodies or extended surfaces, where distances can be measured sometimes to within 0.1 nm, where the forces are large and measurable, and where entropic (thermal) effects are negligible. It is from such experiments (illustrated in Fig. 28) that much hard data on intermolecular and surface interactions has emerged.

(iv)   Particle detachment and peeling experiments (Fig. 28a,b) provide information on interparticle adhesion forces and surface adhesion energies of two solid surfaces in contact (i.e., attractive short-range forces).

(v)   Direct measurement of the force between two macroscopic solid surfaces, usually two curved surfaces, as a function of surface separation can provide the full force law. In such experiments (Fig. 28c) smooth surfaces of glass or mica are commonly used. Both attractive and repulsive forces can be measured from the deflection of a sensitive spring or balance arm, and distances are usually measured using an optical technique. The most accurate of these experiments employ molecularly smooth mica surfaces, where surface separations can be measured to within 0.1 nm and forces to within about 10^{-8} N.

(vi)   Various surface studies such as surface tension and contact angle measurements give information on liquid–liquid and solid–liquid adhesion energies (Fig. 28d).

(vii)   The thickness of free soap films and liquid films adsorbed on surfaces (Fig. 28e,f) can be measured as a function of salt concentration or vapour pressure. Such experiments provide information on the long-range repulsive forces stabilizing such films. Various optical techniques (e.g., reflected intensity or ellipsometry) have been used to measure film thickness to within 0.1 nm.

(viii)   Dynamic interparticle separations in liquids can be measured using light, x-ray, and neutron scattering (Fig. 28g,h). In such experiments the particles can be globular or spherical (e.g., colloidal particles, latex particles, viruses), sheet-like (e.g., clays, lipid bilayers), or rod-like (e.g., DNA). The interparticle forces can be varied externally by changing the osmotic pressure or simply applying a compressive pressure via a semipermeable membrane. Such studies provide information on the long-range repulsive forces between particles in liquids.

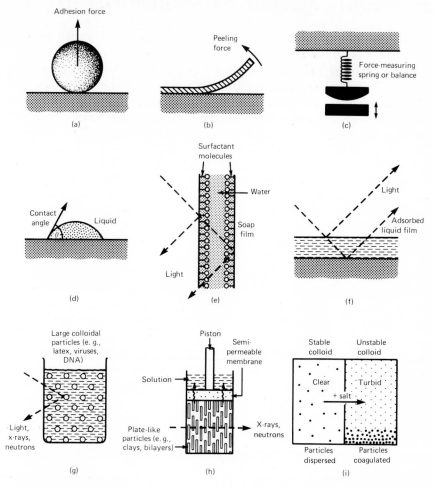

Fig. 28 Different types of measurements that provide information on the forces between particles and surfaces.

(a)   Adhesion measurements (practical applications: xerography, aerosols, crop dusting, particle adhesion).

(b)   Peeling measurements (practical applications: adhesive tapes, crack propagation).

(c)   Direct measurements of force as a function of distance D (practical applications: testing theories of intermolecular forces).

(d)   Contact angle measurements (practical applications: detergency, mineral separation processes using froth flotation, nonstick pans, waterproofed fibres).

(e)   Equilibrium thicknesses of thin free films (practical applications: soap films, foams).

(f)   Equilibrium thickness of thin adsorbed films (practical applications: wetting of hydrophilic surfaces by water, adsorption of molecules from vapour, drainage of liquid layers).

(g)   Interparticle spacing in liquids (practical applications: colloidal suspensions, paints, ink, pharmaceutical dispersions).

(h)   Sheet-like particle spacings in liquids (practical applications: clay and soil swelling behaviour, microstructure of soaps, biological membrane interactions).

(i)   Coagulation studies (practical application: basic experimental technique for testing the stability of colloidal preparations).

(ix) In coagulation studies on colloidal dispersions (Fig. 28i) the salt concentration, pH, or temperature of the suspending liquid medium (usually water) is changed until the dispersion becomes unstable and the particles coalesce (*coagulate* or *flocculate*). Coagulation rates can be very fast or very slow. Such studies provide information on the interplay of repulsive and attractive forces between particles in pure liquids as well as in surfactant and polymer solutions.

Chapter 11

van der Waals Forces between Surfaces

11.1. The force laws for bodies of different geometries: the Hamaker constant

As we saw in Part I, van der Waals forces play a central role in all phenomena involving intermolecular forces, for while they are not as strong as Coulombic or H-bonding interactions, they are always present. When we come to consider the long-range interactions between macroscopic particles and surfaces in liquids we shall find that the two most important forces are the van der Waals and electrostatic forces and that at shorter distances (below 1 to 3 nm) solvation forces often dominate over both.

Let us begin by deriving the van der Waals interaction energies in vacuum for pairs of bodies of different geometries. Starting at the simplest level we shall assume that the interaction is *nonretarded* and *additive*. In Chapter 10 we saw that for an interatomic van der Waals pair potential of the form

$$w = -C/r^6, \qquad (11.1)$$

one may sum (integrate) the energies of all the atoms in one body with all the atoms in the other and thus obtain the "two-body" potential for an atom near a surface [Eq. (10.2)], for a sphere near a surface [Eq. (10.6)], or for two flat surfaces [Eq. (10.8)]. This procedure can be carried out for other geometries as well. The resulting interaction laws for some common geometries are shown in Fig. 29, given in terms of the conventional *Hamaker constant*

$$A = \pi^2 C \rho_1 \rho_1 \qquad (11.2)$$

after Hamaker (1937), who together with Bradley (1932), Derjaguin (1934), and

137

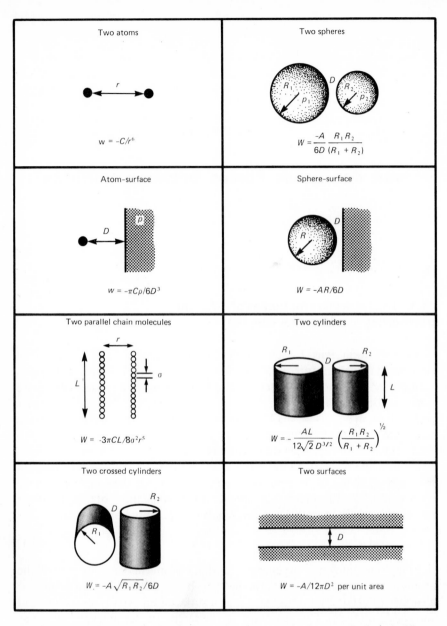

Fig. 29 Nonretarded van der Waals interaction free energies between bodies of different geometries calculated on the basis of pairwise additivity. The *Hamaker constant* A is defined by $A = \pi^2 C \rho_1 \rho_2$, where ρ_1 and ρ_2 are the number of atoms per unit volume in the two bodies and C is the coefficient in the atom–atom pair potential (top left). A more rigorous method of calculating the Hamaker constant in terms of the macroscopic properties of the media is given in Section 11.3. The forces are obtained by differentiating the energies with respect to distance.

de Boer (1936), did much of the earlier work that advanced our understanding of the forces between macroscopic bodies.

Typical values for the Hamaker constants of condensed phases, whether solid or liquid, are about 10^{-19} J *for interactions across vacuum*. For example, if each body is composed of atoms for which $C = 10^{-77}$ J m^6 (cf. Table VIII) and of number density $\rho = 3 \times 10^{28}$ m^{-3} (corresponding to atoms of radius ~ 0.2 nm), the Hamaker constant is

$$A = \pi^2 10^{-77}(3 \times 10^{28})^2 \simeq 10^{-19} \quad \text{J}.$$

Let us consider three cases more specifically. First, for hydrocarbons, treating them as an assembly of CH_2 groups, we have $C \approx 5 \times 10^{-78}$ J m^6 and $\rho = 3.3 \times 10^{28}$ m^{-3} per CH_2 group, from which we obtain $A \approx 5 \times 10^{-20}$ J. This is shown below together with similarly calculated estimates for carbon tetrachloride and water.

Medium	C (10^{-79} J m^6)	ρ (10^{28} m^{-3})	A (10^{-19} J)
Hydrocarbon	50	3.3	0.5
CCl$_4$	1500	0.6	0.5
H$_2$O	140	3.3	1.5

Note that all three Hamaker constants are similar even though the media are composed of molecules differing greatly in polarizability and size. This is because the coefficient C in the interatomic pair potential is roughly proportional to the square of the polarizability α, which in turn is roughly proportional to the volume v of an atom (Section 5.1). And since $\rho \propto 1/v$ we see that $A \propto C\rho^2 \propto \alpha^2\rho^2 \propto v^2/v^2 \propto constant$. This, of course, is a gross oversimplification; however, the Hamaker constants of most condensed phases *are* found to lie in the range $(0.4–4)10^{-19}$ J.

Taking $A = 10^{-19}$ J as a typical value, we can now estimate the strength of the van der Waals interaction between macroscopic bodies in vacuum (or air). Thus, for two spheres of radius $R = 1$ cm $= 10^{-2}$ m in contact at $D \approx 0.2$ nm, their adhesion force will be

$$F = AR/12D^2 = (10^{-19} \times 10^{-2})/12(2 \times 10^{-10})^2$$
$$= 2 \times 10^{-3} \quad \text{N} \approx 0.2 \quad \text{g,}$$

while at $D = 10$ nm the force will have fallen by a factor of 2500 to about 10^{-6} N. At $D = 10$ nm the interaction *energy* is $-AR/12D \approx -10^{-14}$ J, or about $2 \times 10^6 \ kT$, and even for particles with radii as small as $R = 20$ nm their energy exceeds kT at $D = 10$ nm.

For two planar surfaces in contact, the adhesive pressure will be

$$P = A/6\pi D^3 \simeq 7 \times 10^8 \quad \text{N m}^{-2} \approx 7000 \quad \text{atm},$$

while at $D = 10$ nm the pressure is reduced by a factor of about 10^5 to a still-significant 0.05 atm. At contact the energy will be $-A/12\pi D^2 \approx -66$ mJ m^{-2}, of the order expected for the surface energies of solids, discussed later. We see, therefore, that the van der Waals interaction between macroscopic particles is large, and not only when the bodies are in contact.

11.2. The Lifshitz theory of van der Waals forces

The assumptions of simple pairwise additivity inherent in the formulae of Fig. 29 and the definition of A ignore the influence of neighbouring atoms on the interaction between any pair of atoms. First, the polarizability of an atom can change when surrounded by other atoms. Second, recalling our earlier simple model of the dispersion interaction between two Bohr atoms 1 and 2, if a third atom 3 is present, it too will be polarized by the instantaneous field of atom 1, and its induced dipole field will also act on atom 2. Thus, the field from atom 1 reaches atom 2 both directly and by reflection from atom 3. The existence of multiple reflections and the extra force terms to which they give rise is a further instance where straightforward additivity breaks down, and the matter becomes very complicated when many atoms are present. In rarefied media (gases) these effects are small, and the assumptions of additivity hold, but this is not the case for condensed media. Further, the additivity approach cannot be readily extended to bodies interacting in a medium.

The problem of additivity is completely avoided in the *Lifshitz theory* where the atomic structure is ignored and the forces between large bodies, now treated as continuous media, are derived in terms of such bulk properties as their dielectric constants and refractive indices. However, before we proceed it is well to point out that all the expressions in Fig. 29 for the interaction energies remain valid even within the framework of continuum theories. The only thing that changes is the way the Hamaker constant is calculated.

The original Lifshitz theory (Lifshitz, 1956; Dzyaloshinskii *et al.*, 1961) requires a thorough working knowledge of quantum field theory for its understanding, and it is probably due to this fact that it was largely ignored by most scientists who persisted with the additivity approach of Hamaker. Later, Langbein, Ninham, Parsegian, Van Kampen, and others, showed how the main equations could be derived using much simpler theoretical techniques (for reviews see Israelachvili and Tabor, 1973; Parsegian, 1973; Israelachvili

1974; Mahanty and Ninham, 1976). Here we shall adopt the simplest of these using a modified additivity approach. We have already seen that the van der Waals interaction is essentially electrostatic, arising from the dipole field of an atom "reflected back" by a second atom that has been polarized by this field. Let us first analyse this reflected field when the second atom is replaced by a planar surface and where the first atom is replaced by a charge Q (Fig. 30). From Fig. 4 we know that the interaction energy of a charge with a molecule (Fig. 30a) is given by

$$w(r) = -C/r^n = -Q^2\alpha_2/2(4\pi\varepsilon_0\varepsilon_3)^2r^4, \qquad (11.3)$$

where α_2 is the excess polarizability of the molecule 2 in medium 3. When the molecule is replaced by a medium (Fig. 30b) the interaction between the charge in medium 3 and the surface of medium 2 may be obtained by the method of additivity, Eq. (10.1), whence we obtain

$$W(D) = -2\pi C\rho_2/(n-2)(n-3)D^{n-3}$$
$$= -\pi Q^2\rho_2\alpha_2/2(4\pi\varepsilon_0\varepsilon_3)^2D. \qquad (11.4)$$

However, it is well known that a charge Q in a medium of dielectric constant ε_3 at a distance D from the plane surface of a second medium of dielectric constant ε_2 experiences a reflected field from the surface as if it were coming from an "image" charge of strength $Q' = -Q(\varepsilon_2 - \varepsilon_3)/(\varepsilon_2 + \varepsilon_3)$ at a distance

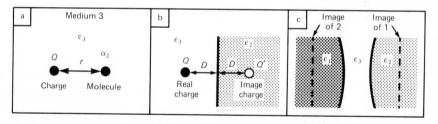

Fig. 30 (a)   A charge interacts with a neutral molecule because of the field reflected by the molecule on becoming polarized.

(b)   Likewise, a charge interacts with a surface because of the field reflected by the surface. This reflected field is the same as if there were an "image" charge Q' at a distance $2D$ from Q. Similarly, a dipole near a surface will see an image of itself reflected by the surface. If $\varepsilon_2 > \varepsilon_3$, the interaction is attractive; if $\varepsilon_2 < \varepsilon_3$, it is repulsive.

(c)   Two surfaces will see an image of each other reflected by each surface that gives rise to the van der Waals force between them. In principle, the reflected or image fields are the same as occur when one looks at a glass surface or a mirror. Metal surfaces reflect most of the light falling on them, and the van der Waals force between metals is much stronger than that between simple dielectric media.

D on the other side of the boundary (Bleaney and Bleaney, 1959, Ch. 2; Landau and Lifshitz, 1963, Ch. II). Accordingly, the interaction energy is also given by

$$W(D) = -\frac{Q^2}{4(4\pi\varepsilon_0\varepsilon_3)D}\left(\frac{\varepsilon_2 - \varepsilon_3}{\varepsilon_2 + \varepsilon_3}\right). \tag{11.5}$$

Comparing the above two equations we immediately find that

$$\rho_2\alpha_2 = 2\varepsilon_0\varepsilon_3(\varepsilon_2 - \varepsilon_3)/(\varepsilon_2 + \varepsilon_3). \tag{11.6}$$

This is an important result, giving the excess bulk or volume polarizability of a dielectric medium 2 in medium 3 in terms of the purely macroscopic properties of the media. The Hamaker constant for the interaction of two media 1 and 2 across a third medium 3 (Fig. 30c) may now be expressed in terms of Eq. (6.18) for C and Eq. (11.6) for $\rho_1\alpha_1$ and $\rho_2\alpha_2$ as follows:

$$A = \pi^2 C \rho_1 \rho_2$$

$$= \frac{6\pi^2 kT\rho_1\rho_2}{(4\pi\varepsilon_0)^2} \sum_{n=0,1,\dots}^{\infty}{}' \frac{\alpha_1(iv_n)\alpha_2(iv_n)}{\varepsilon_3^2(iv_n)}$$

$$= \frac{3}{2}kT \sum_{n=0,1,\dots}^{\infty}{}' \left[\frac{\varepsilon_1(iv_n) - \varepsilon_3(iv_n)}{\varepsilon_1(iv_n) + \varepsilon_3(iv_n)}\right]\left[\frac{\varepsilon_2(iv_n) - \varepsilon_3(iv_n)}{\varepsilon_2(iv_n) + \varepsilon_3(iv_n)}\right]. \tag{11.7}$$

Replacing the sum by the integral of Eq. (6.25) we end up with

$$A \simeq \frac{3}{4}kT\left(\frac{\varepsilon_1 - \varepsilon_3}{\varepsilon_1 + \varepsilon_3}\right)\left(\frac{\varepsilon_2 - \varepsilon_3}{\varepsilon_2 + \varepsilon_3}\right)$$

$$+ \frac{3h}{4\pi}\int_{v_1}^{\infty}\left(\frac{\varepsilon_1(iv) - \varepsilon_3(iv)}{\varepsilon_1(iv) + \varepsilon_3(iv)}\right)\left(\frac{\varepsilon_2(iv) - \varepsilon_3(iv)}{\varepsilon_2(iv) + \varepsilon_3(iv)}\right)dv, \tag{11.8}$$

where ε_1, ε_2, and ε_3 are the static dielectric constants of the three media, $\varepsilon(iv)$ are the values of ε at imaginary frequencies, and $v_n = (2\pi kT/h)n = 4 \times 10^{13}n$ s^{-1} at 300 K. The first term in Eq. (11.8) gives the zero-frequency energy of the van der Waals interaction and includes the Keesom and Debye contributions. The second term gives the dispersion energy and includes the London energy contribution. Equations (11.7) and (11.8) are not exact but are only the first terms in an infinite series for the nonretarded Hamaker constant. The other terms, however, are small and rarely contribute more than 5%.

The above approach may also be used to calculate the van der Waals interaction between a molecule or small particle 1 in medium 3 with the surface

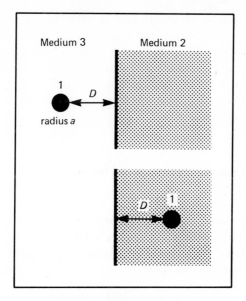

Fig. 31

of medium 2 (Fig. 31a). The interaction energy may be readily shown to be given by (Israelachvili, 1974)

$$W(D) = -\pi C \rho_2 / 6D^3$$

$$= -\frac{kTa_1^3}{2D^3} \sum{}' \left[\frac{\varepsilon_1(iv_n) - \varepsilon_3(iv_n)}{\varepsilon_1(iv_n) + 2\varepsilon_3(iv_n)} \right] \left[\frac{\varepsilon_2(iv_n) - \varepsilon_3(iv_n)}{\varepsilon_2(iv_n) + \varepsilon_3(iv_n)} \right]$$

$$\approx -\frac{Aa_1^3}{3D^3}. \tag{11.9}$$

If, on the other hand, molecule 1 is in medium 2 (Fig. 31b), we must exchange $\varepsilon_1(iv)$ and $\varepsilon_3(iv)$ in this equation. Equation (11.9) has interesting consequences when the dielectric media 2 and 3 are liquids, or at least permeable to molecule 1, for it predicts that the molecule will behave in one of two ways depending on the relative values of the dielectric permittivities:

(i)   The molecule will be attracted to the interface from either side of it, (e.g., if ε_1 is intermediate between ε_2 and ε_3).

(ii)   The molecule will be attracted towards the interface and then repelled from the other side (e.g., if $\varepsilon_1 > \varepsilon_3 > \varepsilon_2$, the molecule will be driven from right to left in Fig. 31).

Thus, in the absence of constraints or other forces the van der Waals interaction alone may promote the migration of small uncharged particles across liquid interfaces. Note that repulsion from both sides of the interface never occurs. These effects are entirely analogous to those discussed in Section 9.4 in nonspecific terms. The application of the Lifshitz theory to explain the engulfing or rejection of particles by surfaces has been extensively studied by Van Oss et al. (1980).

11.3. Hamaker constants calculated on the basis of the Lifshitz theory

The dielectric permittivity $\varepsilon(iv)$ of a medium varies with frequency v in much the same way as does the atomic polarizability of an atom, discussed in Section 6.5. Thus, $\varepsilon(v)$ and $\varepsilon(iv)$ can usually be represented by a function of the form (Mahanty and Ninham, 1976)

$$\varepsilon(v) = 1 + \frac{(\text{constant})}{(1 - iv/v_{\text{rot}})} + \frac{(\text{constant})}{(1 - v^2/v_{\text{e}}^2)}, \tag{11.10}$$

so that

$$\varepsilon(iv) = 1 + \frac{(\varepsilon - n^2)}{(1 + v/v_{\text{rot}})} + \frac{(n^2 - 1)}{(1 + v^2/v_{\text{e}}^2)}, \tag{11.11}$$

$$\varepsilon(0) = \varepsilon,$$

where v_{rot} is the rotational relaxation frequency typically at microwave and lower frequencies ($v_{\text{rot}} < 10^{12}$ s^{-1}), v_{e} the main electronic absorption frequency in the UV typically around 3×10^{15} s^{-1}, and n the refractive index of the medium in the visible [i.e., $n^2 = \varepsilon_{\text{vis}}(v)$]. Since $v_1 \approx 4 \times 10^{13}$ s$^{-1} \gg v_{\text{rot}}$ the dispersion energy is determined solely by the electronic absorption [third term in Eq. (11.11)]. We may therefore substitute an expression of the form

$$\varepsilon(iv) = 1 + (n^2 - 1)/(1 + v^2/v_{\text{e}}^2) \tag{11.12}$$

for each medium into Eq. (11.8) and integrate using the definite integral of Eq. (6.27). If the absorption frequencies of all three media are assumed to be at the same v_{e}, we obtain the following approximate expression for the non-retarded dispersion contribution to the Hamaker constant:

$$A_{v>0} \approx \frac{3hv_{\text{e}}}{8\sqrt{2}} \frac{(n_1^2 - n_3^2)(n_2^2 - n_3^2)}{(n_1^2 + n_3^2)^{1/2}(n_2^2 + n_3^2)^{1/2}\{(n_1^2 + n_3^2)^{1/2} + (n_2^2 + n_3^2)^{1/2}\}}, \tag{11.13a}$$

which together with the zero-frequency part

$$A_{v=0} = \frac{3}{4}kT\left(\frac{\varepsilon_1 - \varepsilon_3}{\varepsilon_1 + \varepsilon_3}\right)\left(\frac{\varepsilon_2 - \varepsilon_3}{\varepsilon_2 + \varepsilon_3}\right) \tag{11.13b}$$

gives the total Hamaker constant for two macroscopic phases 1 and 2 interacting across a medium 3. For two identical phases 1 interacting across medium 3, we obtain

$$A = A_{v=0} + A_{v>0} = \frac{3}{4}kT\left(\frac{\varepsilon_1 - \varepsilon_3}{\varepsilon_1 + \varepsilon_3}\right)^2 + \frac{3hv_e}{16\sqrt{2}}\frac{(n_1^2 - n_3^2)^2}{(n_1^2 + n_3^2)^{3/2}}. \tag{11.14}$$

For interactions in vacuum, $\varepsilon_3 = 1$ and $n_3 = 1$ in the above equations. Four interesting aspects of the above equation may be noted:

(1)   The van der Waals force between any two bodies in vacuum is always attractive.

(2)   The van der Waals force between two identical bodies in a medium is always attractive (A positive), while that between different bodies in a medium can be attractive or repulsive (A negative).

(3)   The Hamaker constant for two similar media interacting across another medium remains unchanged if the media are interchanged. Thus, if no other forces are operating, a liquid film in air will always tend to thin under the influence of the attractive van der Waals force between the two surfaces, or in other words, two air phases appear to attract each other across a thin film.

(4)   The purely entropic zero-frequency contribution $A_{v=0}$ can never exceed $\frac{3}{4}kT$, or about 3×10^{-21} J at 300 K. For interactions across vacuum where the dispersion energy contribution $A_{v>0}$ is typically $\sim 10^{-19}$ J, this is always small; but as will be seen for interactions across a medium of high ε it can dominate over the dispersion energy contribution.

A number of authors have given more complex analytic formulae for A when the interacting media have different absorption frequencies (Israelachvili, 1972; Horn and Israelachvili, 1981, Appendix B), while others have described simple numerical procedures for computing Hamaker constants for media with many absorption frequencies when the more exact equation (11.7) must be used (Gregory, 1970; Pashley, 1977; Hough and White, 1980).

For interactions involving conducting media (i.e., metals), the dispersion formula of Eq. (11.11) does not apply. The dielectric permittivity of a metal is given approximately by

$$\varepsilon(v) = 1 - v_e^2/v^2, \tag{11.15}$$

TABLE XV

Nonretarded Hamaker constants for two identical media interacting across vacuum (air)

Medium	Dielectric Constant ε	Refractive Index n	Absorption Frequency[a] ν_e $(10^{15}\ s^{-1})$	Hamaker Constant A $(10^{-20}\ J)$		
				Eq. (11.14) $\varepsilon_3 = 1$	Exact Solutions[b]	Experimental[c]
Water	80	1.333	3.0	3.7	*3.7, 4.0*	
n-Pentane	1.84	1.349	3.0	3.8	3.75	
n-Octane	1.95	1.387	3.0	4.5	4.5	
n-Dodecane	2.01	1.411	3.0	5.0	5.0	
n-Tetradecane	2.03	1.418	2.9	5.0	5.1, 5.4	
n-Hexadecane	2.05	1.423	2.9	5.1	5.2	
Hydrocarbon (cryst.)	2.25	1.50	3.0	7.1		10
Cyclohexane	2.03	1.426	2.9	5.2		
Benzene	2.28	1.501	2.1	5.0		
Carbon tetrachloride	2.24	1.460	2.7	5.5		
Acetone	21	1.359	2.9	4.1		
Ethanol	26	1.361	3.0	4.2		
Polystyrene	2.55	1.557	2.3	6.5	*6.6, 7.9*	
Polyvinyl chloride	3.2	1.527	2.9	7.5	7.8	
PTFE	2.1	1.359	2.9	3.8	3.8	
Fused quartz	3.8	1.448	3.2	6.3	6.5	5–6
Mica	7.0	1.60	3.0	10	10	13.5
CaF$_2$	7.4	1.427	3.8	7.0	7.2	
Liquid He	1.057	1.028	5.9	0.057		
Metals (Au, Ag, Cu)	∞	—	3–5	25–40	*30–50*	
				Eq. (11.17)		

[a] UV absorption frequencies obtained from "Cauchy plots," mainly from Hough and White (1980) and H. Christenson (1983, thesis).
[b] Exact solutions computed mainly by Hough and White (1980) and, when in italics, by Parsegian and Weiss (1981).

so that

$$\varepsilon(iv) = 1 + v_e^2/v^2, \qquad (11.16)$$

where v_e is the so called plasma frequency of the free electron gas, typically in the range $(3-5) \times 10^{15}$ s^{-1}. Substituting the above into Eq. (11.8) and integrating as before we obtain for two metals interacting across vacuum

$$A \approx (3/16\sqrt{2})hv_e. \qquad (11.17)$$

Table XV gives some Hamaker constants for two identical phases interacting across vacuum as calculated using Eq. (11.14) together with more rigorously computed values. Experimentally determined values are also included where available. Note the good agreement between values calculated on the basis of Eq. (11.14) and those computed more rigorously by solving the full Lifshitz equation. Note too the reasonably good agreement between theory and experiment. We may also compare the values of A of Table XV with those obtained from a summation of additive pair potentials discussed in Section 11.1. Thus, for hydrocarbons and CCl$_4$, the agreement is surprisingly good, but for highly polar liquids such as water, the additivity approach overestimates the value of A mainly because it overestimates the zero-frequency contribution.

11.4. Applications of the Lifshitz theory to interactions in a medium

The Lifshitz theory is particularly suitable for analysing the interactions of different phases across a medium. As a graphic example, Fig. 32 shows how $\varepsilon(iv)$ varies with frequency for water and a typical hydrocarbon such as dodecane, for both of which absorption frequencies are about $v_e \approx 3.0 \times 10^{15}$ s^{-1}. The area under the dashed curve at frequencies above v_1 is roughly proportional to the nonretarded dispersion energy of two hydrocarbon phases across a water film, for which we obtain, using Eq. (11.14),

$$A_{v>0} \approx \frac{3(6.63 \times 10^{-34})(3 \times 10^{15})}{16\sqrt{2}} \frac{(1.41^2 - 1.33^2)^2}{(1.41^2 + 1.33^2)^{3/2}} \approx 0.17 \times 10^{-20} \quad \text{J}.$$

Concerning the zero-frequency contribution, note that water exhibits strong absorptions at lower frequencies and consequently has a high static dielectric constant of $\varepsilon \approx 80$. In contrast, the dielectric constant of hydrocarbon remains the same right down to zero frequency where $\varepsilon \approx n^2 \approx 2.0$. The large difference between ε_{H_2O} and ε_{hc} leads to a large zero-frequency

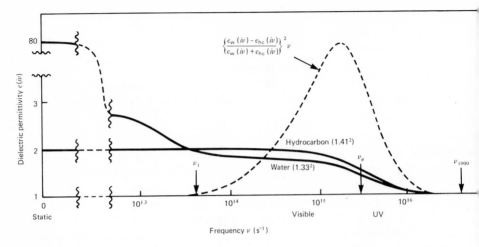

Fig. 32 Dielectric permittivity $\varepsilon(iv)$ as a function of frequency v for water and hydrocarbon. In the visible and UV range these are given by

$$\varepsilon_w(iv) = 1 + (n_w^2 - 1)/(1 + v^2/v_e^2), \qquad n_w = 1.33,$$

$$\varepsilon_{hc}(iv) = 1 + (n_{hc}^2 - 1)/(1 + v^2/v_e^2), \qquad n_{hc} = 1.41,$$

where $v_e = 3.0 \times 10^{15} \text{ s}^{-1}$ for both. The total Hamaker constant for this system at 300 K is about 0.45×10^{-20} J.

contribution to the Hamaker constant of

$$A_{v=0} = \frac{3}{4}kT\left(\frac{80 - 2}{80 + 2}\right)^2 \approx 0.28 \times 10^{-20} \quad \text{J at 300 K,}$$

giving a total value for A of $(0.17 + 0.28)10^{-20} \approx 0.45 \times 10^{-20}$ J.

Ninham and Parsegian (1970), Gingell and Parsegian (1972), and more recently Hough and White (1980) considered the hydrocarbon–water system in great detail and found that the theoretical A for this system lies in the range $(4–7) \times 10^{-21}$ J depending on the refractive index of the hydrocarbon. About half the interaction energy comes from the purely entropic zero-frequency contribution—a situation arising from the unusually high static dielectric constant of water and the similarity in the refractive indices of hydrocarbons and water. Note that as far as van der Waals forces are concerned the interaction of hydrocarbon across water is the same as for water across hydrocarbon. Experimental A values for hydrocarbon–water systems have been determined from studies on lipid bilayers. Table XVI gives some computed values of A for interactions across various media based on the

TABLE XVI
Hamaker constants for two media interacting across another medium

Interacting Media			Hamaker Constant A (10^{-20} J)		
				Exact	
1	3	2	Eq. (11.13)[a]	Solutions[b]	Experimental
Air	Water	Air	3.7	3.70	
Pentane	Water	Pentane	0.28	0.34	
Octane	Water	Octane	0.36	0.41	
Dodecane	Water	Dodecane	0.44	0.50	0.5[d]
Hexadecane	Water	Hexadecane	0.49	0.54	0.3–0.6[d,e]
Water	Hydrocarbon	Water	0.3–0.5	0.34–0.54	0.3–0.9[f]
Polystyrene	Water	Polystyrene	1.4	0.95, 1.3[c]	
Fused quartz	Water	Fused quartz	0.63	0.83	
Fused quartz	Octane	Fused quartz	0.13		
PTFE	Water	PTFE	0.29	0.33	
Mica	Water	Mica	2.0	2.0	2.2[g]
Ag, Au, Cu	Water	Ag, Au, Cu	–	30–40[c]	40 (Au)[h]
Water	Pentane	Air	0.08	0.11	
Water	Octane	Air	0.51	0.53	
Octane	Water	Air	−0.24	−0.20	
Fused quartz	Water	Air	−0.87	−1.0	
Fused quartz	Octane	Air	−0.7		
Fused quartz	Tetradecane	Air	−0.4		−0.5[h]
CaF$_2$, SrF$_2$	Liquid He	Vapour	−0.59	−0.59[i]	−0.58[i]

Based on dielectric data of Table XV, assuming mean values for v_e.
Hough and White (1980). [c] Parsegian and Weiss (1981). [d] Lis et al. (1982).
Ohshima et al. (1982). [f] Requena et al. (1975). [g] Israelachvili and Adams (1978).
Derjaguin et al. (1978). [i] Sabisky and Anderson (1973).

Lifshitz theory, together with measured values where these are available. As can be seen, the agreement between theory and experiment is very good. Note the much smaller Hamaker constants compared to those for interactions across vacuum (Table XV).

11.5. Repulsive van der Waals forces

We have already seen in Chapter 6 how repulsive van der Waals forces can arise in a medium and again in Chapter 9 how repulsive forces in general can be understood intuitively. Equation (11.13) shows that the Hamaker constant will be negative, resulting in repulsion, whenever the dielectric properties of the intervening medium are intermediate between those of the two interacting media. Indeed, one of the early successes of the Lifshitz theory was in the quantitative explanation of the wetting properties of liquid helium due to a negative Hamaker constant (Dzyaloshinskii et al., 1961). It is well known that liquid helium avidly spreads on almost any surface with which it comes into contact. Thus, if liquid helium is placed in a beaker, it rapidly climbs up the walls and down the other side, and eventually leaves the beaker altogether. The reason for this peculiar behaviour is that the dielectric permittivity of liquid helium, $\varepsilon = n^2 = 1.057$, is lower than that of any other medium (except vapour). Thus, there will be a negative Hamaker constant and a repulsive van der Waals force across an adsorbed liquid He film which will act to thicken the film so as to lower its energy. When liquid He climbs up a smooth wall the gain in van der Waals energy, however, is at the expense of gravitational energy, and so the equilibrium film thickness will decrease with height (Fig. 33). Sabisky and Anderson (1973) measured the thickness D as a function of the height H of liquid helium films at 1.38 K on atomically smooth surfaces of CaF_2, SrF_2, and BaF_2. In the range of H from 1 to 10^4 cm D decreased from ~ 25 to ~ 1 nm. Let us look at this interesting phenomenon of *wetting films* in more detail (see also Section 12.8).

The relationship between the height of the film H and the film thickness D can be obtained by equating the chemical potentials of all the helium atoms as follows. First, the change in energy per helium atom added to the film is $\partial W/\partial N$ where $W(D) = A/12\pi D^2$ per unit area. Now if v is the volume occupied per He atom, then $N = D/v$, and so we have

$$\frac{\partial W}{\partial N} = \frac{\partial W}{\partial D}\frac{\partial D}{\partial N} = P(D)v = \frac{Av}{6\pi D^3},$$

where the negative pressure $P(D)$ is often referred to as the *disjoining pressure* of a film. The condition of equilibrium throughout the film is therefore

$$\mu = P(D)v + mgH = \frac{Av}{6\pi D^3} + mgH = \text{constant} = 0, \qquad (11.18)$$

since at $H = 0$, $D = \infty$. From Table XVI we have $A = -5.9 \times 10^{-21}$ J, and for helium $v = (0.36 \text{ nm})^3 = 4.7 \times 10^{-29}$ m^3 and $mg = 6.55 \times 10^{-26}$ J m^{-1}. Thus, we expect for the film profile

$$D = \frac{28}{\sqrt[3]{H(\text{cm})}} \quad \text{nm.}$$

Sabisky and Anderson (1973) measured $D = 2.8$ nm at $H = 1000$ cm, exactly as expected. But at $H = 1$ cm they obtained $D = 21.5$ nm instead of $D = 28$ nm due to retardation effects, which were observed for films thicker than 6 nm.

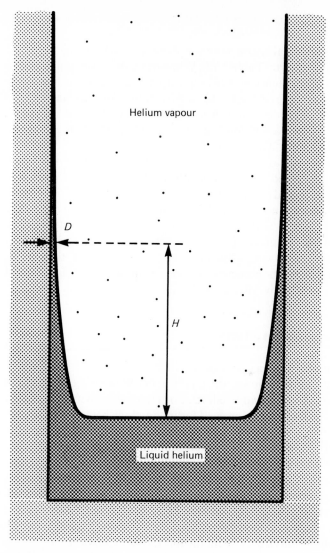

Fig. 33 Liquid helium climbs up the walls of containers because of the repulsive van der Waals force across the adsorbed film. At equilibrium the work done against gravity mgH is balanced by the work done by the van der Waals force.

Repulsive van der Waals forces also occur across thin alkane films on alumina. The measured variation of the disjoining pressure with thickness in the range 0.5–80 nm has been found to be in excellent agreement with theory

(Blake, 1975). Once again retardation effects are evident at film thicknesses beyond 5 nm.

Negative Hamaker constants are also expected for water films on hydrocarbons and on quartz (Table XVI). While water does wet quartz, this is also due in part to the existence of other repulsive forces across the film such as electric double-layer forces (Chapter 12) and hydration forces (Chapter 13). However, water certainly does not wet hydrocarbons because of the attractive *hydrophobic* solvation force across the film, which is stronger than the repulsive van der Waals force (Chapter 13).

Repulsive van der Waals forces often occur between different organic polymers dissolved in an organic solvent. This is because the refractive indices of such compounds are always very similar. Van Oss *et al.* (1980) estimated the Hamaker constants of 31 polymer pairs in different solvents. These varied from $A = +0.75 \times 10^{-20}$ J to $A = -0.38 \times 10^{-20}$ J. It was found that whenever the Hamaker constant exceeds 0.03×10^{-20} J the polymers are miscible in the solvent, while for $A < -0.03 \times 10^{-20}$ J, they are immiscible, and phase separation occurs. Another example of repulsive van der Waals forces is given below after a few words about retardation effects.

11.6. Retardation effects

In Chapter 6 it was pointed out that at distances beyond about 5 nm the dispersion contribution to the total van der Waals force begins to decay more rapidly due to retardation effects. For interactions between molecules, this is of little consequence. However, for interactions between macroscopic bodies in a medium, where the forces can still be important at such large distances, the effects of retardation must sometimes be taken into account.

There is no simple equation for calculating the van der Waals force at *all* separations. Strictly, the Hamaker constant is never truly constant but decreases progressively as D increases from zero, and it can fall to less than half its nonretarded value (i.e., at $D = 0$) by $D = 10$ nm. Mahanty and Ninham (1976) described numerical methods for computing the van der Waals force law at all distances by solving the full Lifshitz equation.

As an illustration of retardation effects, Fig. 34 shows experimental results obtained for the van der Waals force law between two curved mica surfaces in aqueous electrolyte solutions (in a liquid medium such as water the full force law also involves the electric double-layer force, so that some extrapolation and subtraction of this force must be done in order to obtain the purely van der Waals force contribution to the total interaction). For this system, the nonretarded Hamaker constant was found to be $A = 2.2 \times 10^{-20}$ J. The results clearly show the onset of retardation at separations above about

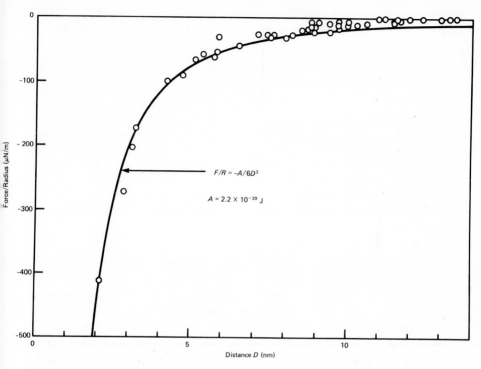

Fig. 34 Attractive van der Waals force F between two curved mica surfaces of radius $R \approx 1$ cm measured in water and aqueous electrolyte solutions. The measured nonretarded Hamaker constant is $A = 2.2 \times 10^{-20}$ J. Retardation effects are apparent at distances above 5 nm. [From Israelachvili and Adams (1978) and Israelachvili and Pashley (unpublished).]

5 nm—the same distance as in the wetting film experiments described above. Retarded van der Waals forces have also been measured between glass and metal surfaces in air or vacuum out to distances of 1200 nm, all in good agreement with the Lifshitz theory (see Section 11.9). Indeed, the first direct measurements of van der Waals forces by Derjaguin and Abrikossova (1954) were of retarded forces between two quartz glass surfaces in the distance range 100–400 nm.

Since only the dispersion force suffers retardation, and not the zero-frequency term,* some very interesting effects can sometimes arise leading to a change in sign of the Hamaker constant at some finite separation. A particularly notable example of this phenomenon concerns the behaviour of

---

* However, the zero-frequency energy is *screened* by electrolyte ions in water, and $A_{\nu=0}$ decays roughly exponentially with distance with a decay length equal to half the *Debye length* (see Section 17.1).

Wetting film   Nonwetting film

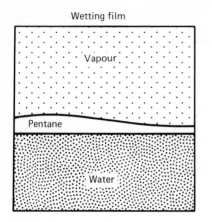 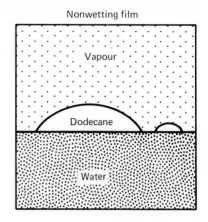

Fig. 35

liquid hydrocarbons on water. Referring to Table XVI we note that the nonretarded Hamaker constant for a pentane film on water is very small, about $A \approx 10^{-21}$ J. This is made up from a zero-frequency contribution of $A_{v=0} \approx -0.8 \times 10^{-21}$ J together with a dispersion contribution of $A_{v>0} \approx 1.6 \times 10^{-21}$ J. At larger distances, however, $A_{v>0}$ will become retarded and progressively decrease. Thus, at some finite distance the value of A will change sign from net positive to net negative once $A_{v=0}$ begins to dominate the interaction. It is for this reason that pentane spreads on water: the van der Waals force across the film is repulsive, so tending to increase its thickness and thus favouring the spreading of the liquid on water. For the higher alkanes, the attractive dispersion force contribution is very much larger (cf. $A_{v>0} \approx 6 \times 10^{-21}$ J for octane), and these hydrocarbons do not spread on water but collect as isolated lenses on the water surface (Fig. 35).

11.7. Combining relations

Combining relations (or combining laws) are frequently used for obtaining approximate values for unknown Hamaker constants in terms of known ones. Let us define A_{132} as the nonretarded Hamaker constant for media 1 and 2 interacting across medium 3 (Fig. 30c). A glance at Eq. (11.7) shows that we may expect A_{132} to be approximately related to A_{131} and A_{232} via

$$A_{132} \approx \pm \sqrt{A_{131} A_{232}} \tag{11.19}$$

or

$$A_{12} \approx \sqrt{A_{11} A_{22}}, \tag{11.20}$$

where A_{12} is for media 1 and 2 interacting across vacuum (i.e., with no medium 3 between them). This is a useful combining relation giving A_{12} in terms of the Hamaker constants of the individual media. Two other useful relations are (Israelachvili, 1972)

$$A_{131} = A_{313} \approx A_{11} + A_{33} - 2A_{13} \tag{11.21}$$

$$\approx (\sqrt{A_{11}} - \sqrt{A_{33}})^2, \tag{11.22}$$

which when combined with Eq. (11.19) above gives

$$A_{132} \approx (\sqrt{A_{11}} - \sqrt{A_{33}})(\sqrt{A_{22}} - \sqrt{A_{33}}). \tag{11.23}$$

Note the similarity of these combining relations to those derived in Chapter 9, Eqs. (9.1)–(9.9).

As an illustration of the above relations let us consider a few systems whose Hamaker constants are given in Tables XV and XVI. Thus, for the quartz–octane–air system, Eq. (11.23) would predict for A_{132}:

$$A_{132} \approx (\sqrt{6.3} - \sqrt{4.5})(0 - \sqrt{4.5})10^{-20} = -0.82 \times 10^{-20} \quad \text{J},$$

to be compared with the more rigorously computed value of -0.71×10^{-20} J. Likewise for the CaF_2–helium–vapour system, we expect

$$A_{132} \approx (\sqrt{7.2} - \sqrt{0.057})(0 - \sqrt{0.057})10^{-20} = -0.58 \times 10^{-20} \quad \text{J}$$

to be compared with -0.59×10^{-20} J.

For the system quartz–octane–quartz, Eq. (11.23) gives

$$A_{131} \approx (\sqrt{6.3} - \sqrt{4.5})^2 10^{-20} = 0.15 \times 10^{-20} \quad \text{J}$$

compared to 0.13×10^{-20} J.

Combining relations are applicable only when dispersion forces dominate the interactions as in the above examples, but they break down when applied to media with high dielectric constants such as water whenever the zero-frequency contribution $A_{v=0}$ is large. Thus, for hexadecane across water, Eq. (11.23) would predict a Hamaker constant of $(\sqrt{5.2} - \sqrt{3.7})^2 10^{-20} \approx 0.13 \times 10^{-20}$ J, which is much smaller than the real value of $\sim 0.5 \times 10^{-20}$ J.

In view of the ease with which Hamaker constants may be reliably computed, using Eq. (11.13) or more rigorous numerical methods, the use of Combining relations is not recommended.

11.8. Surface energies

Chapter 14 describes various phenomena that arise from the *surface energies* γ of solids and liquids (for a liquid, γ is usually referred to as its *surface tension*). Here we shall see how surface energies are determined from the forces between molecules and surfaces. But first, what do we mean by the surface energy of a material? Let us go back to Fig. 25 and the pairwise summation of energies between the atoms of one medium with all the atoms of the *other* medium, which, for van der Waals forces, gave us the interaction energy between two identical media as $W = -A/12\pi D^2$. Had the summation been carried out between all atoms, including atoms in the *same* medium, we should have obtained two additional energy terms:

$$W = -\text{constant} + A/12\pi D_0^2, \tag{11.24}$$

where the constant is simply the bulk cohesive energy, Eq. (6.5), of the atoms with their immediate neighbours at $D = D_0$ and where the second (positive) term arises from the existence of the two surfaces whose atoms have fewer neighbours than those in the bulk. The second term in Eq. (11.24) gives the *surface energies* 2γ of the two media, and as can be seen, it arises from unsaturated surface "bonds." A free liquid will always tend to minimize its surface energy (and hence its total energy) by minimizing its surface area.

Thus, apart from the bulk energy the total energy of two planar surfaces at a distance D apart (Fig. 36) may be written as

$$W = \frac{A}{12\pi}\left(\frac{1}{D_0^2} - \frac{1}{D^2}\right) \quad \text{per unit area.} \tag{11.25}$$

At $D = D_0$ (two surfaces in contact), $W = 0$, while for $D = \infty$ (two isolated surfaces),

$$W = A/12\pi D_0^2 = 2\gamma \tag{11.26}$$

or

$$\gamma = A/24\pi D_0^2. \tag{11.27}$$

In other words γ equals half the (van der Waals) energy needed to separate two surfaces from contact to infinity.

We may now test Eq. (11.27) to see how well it predicts the surface energies of materials. Unfortunately it is not at all clear how the interfacial contact separation D_0 is to be determined. At first sight one might expect that D_0 should be the same as the distance between atomic centres and thus be put equal to σ. However, in the process of computing Hamaker constants the surface atoms were artificially "smeared out" by transferring the energy sum to an integral, and this complicates use of Eq. (11.27) at small interatomic distances (Tabor, 1982). Let us first consider the matter in terms of pairwise additivity: For an idealized planar close packed solid, each surface atom (of diameter σ) will have only 9 nearest neighbours instead of 12. Thus, when it comes into contact with a second surface each surface atom will gain $3w = 3C/\sigma^6$ in binding energy. Each atom occupies an area of $\sigma^2 \sin 60°$, and

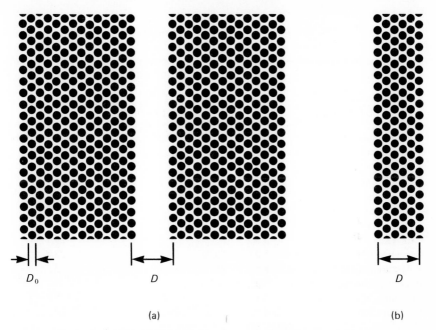

(a)   (b)

Fig. 36 (a)   Two planar media. The pairwise summation of London dispersion energies between *all* atoms gives $W = A/12\pi D_0^2 - A/12\pi D^2$ per unit area, where D_0 is the interfacial contact separation. The first term may be equated with the surface energy $\gamma = A/24\pi D_0^2$ of an *isolated* surface. For two surfaces close together, their total surface energy may therefore be written as $2\gamma(1 - D_0^2/D^2)$. The long-range van der Waals interaction energy can be seen to be no more than a perturbation of the surface energy γ. A similar result is obtained for (b), a thin film.

for a close-packed solid, the number density is $\rho = \sqrt{2}/\sigma^3$. Thus, the surface energy should be approximately

$$\gamma \approx \frac{1}{2}\left(\frac{3w}{\sigma^2 \sin 60°}\right) \approx \frac{3C\rho^2}{4\sigma^2 \sin 60°} \approx \frac{A}{24\pi(\sigma/2.5)^2},$$ (11.28)

where $A = \pi^2 C\rho^2$ is the Hamaker constant, Eq. (11.2).

Thus, to use Eq. (11.27) for calculating surface energies we must use a "cut-off" distance D_0 that is substantially less than the interatomic or intermolecular centre-to-centre distance σ (Israelachvili, 1973; Aveyard and Saleem, 1976; Hough and White, 1980; Tabor, 1982). Thus, for hydrocarbons where $\sigma \approx 0.4$ nm, we would use $D_0 \approx \sigma/2.5 \approx 0.16$ nm. Table XVII gives the predicted surface energies of a variety of compounds all based on the same cut-off separation of $D_0 = 0.165$ nm. It is remarkable that this "universal constant" for D_0 yields values for γ in such good agreement with those

TABLE XVII

Comparison of experimental surface energies with those calculated on the basis of the Lifshitz theory

Material (ε) in Order of Increasing ε	Theoretical A (10^{-20} J)	Surface Energy γ (mJ m^{-2})	
		$A/24\pi D_0^2$ ($D_0 = 0.165$ nm)	Experimental[a] (20°C)
Liquid helium (1.057)	0.057	0.28	0.12–0.35 (4–1.6 K)
n-Pentane (1.8)	3.75	18.3	16.1
n-Octane (1.9)	4.5	21.9	21.6
Cyclohexane (2.0)	5.2	25.3	25.5
n-Dodecane (2.0)	5.0	24.4	25.4
n-Hexadecane (2.1)	5.2	25.3	27.5
PTFE (2.1)	3.8	18.5	18.3
CCl$_4$ (2.2)	5.5	26.8	29.7
Benzene (2.3)	5.0	24.4	28.8
Polystyrene (2.6)	6.6	32.1	33
Polyvinyl chloride (3.2)	7.8	38.0	39
Acetone (21)	4.1	20.0	23.7
Ethanol (26)	4.2	20.5	22.8
Methanol (33)	3.6	18	23
Glycol (37)	5.6	28	48
Glycerol (43)	6.7	33	63
Water (80)	3.7	18	73
H$_2$O$_2$ (84)	5.4	26	76
Formamide (109)	6.1	30	58

[a] Note the good agreement between theory and experiment (within 20%) except for the last six strongly H-bonding liquids.

measured, even for very different liquids and solids. Only for highly polar H-bonding liquids (methanol, glycol, water, etc.) does Eq. (11.27) seriously underestimate their surface energies, which is to be expected (Section 8.4). But for "ordinary" solids and liquids, even including acetone and ethanol, Eq. (11.27) appears to be reliable to within 10 to 20% (which is anyway within the accuracy that A can be computed). We may therefore conclude with the following simple formula for estimating the Hamaker constants of non–H-bonding solids and liquids from their surface energies

$$\gamma \approx A/24\pi(0.165 \text{ nm})^2, \qquad (11.29)$$

which leads to

$$A \approx 2.1 \times 10^{-21} \gamma, \qquad (11.30)$$

where γ is in millijoules per square meter and A in joules.

11.9. Experiments on van der Waals forces

The expression "experiments on van der Waals forces" is a rather vague one since many important phenomena in physics and chemistry involve van der Waals interactions in one way or another. Thus, the measurement of the surface tensions of nonpolar liquids or the thickness of wetting films may be thought of as experiments on van der Waals forces, and we have seen how the results are in good agreement with theory. Other phenomena involving van der Waals interactions (e.g., physical adsorption, adhesion, the strength of solids) have also been extensively studied; but experiments of this type afford a poor way of studying van der Waals forces since they have usually to contend with many parameters that themselves are uncertain, so that it becomes difficult to compare the results with theory.

The most direct way to study van der Waals forces is simply to position two bodies close to each other and to measure the force of attraction as a function of the distance between them. The pioneering measurements carried out by Derjaguin's school in the USSR and Overbeek's school in The Netherlands in the 1950s and 1960s were of this type; the bodies were made of glass, the force was determined by measuring the deflection of a sensitive spring or balance arm, and the distance between the highly polished surfaces was obtained by optical interference (using Newton's rings, for example). In this way the forces between various types of glass in air or vacuum were successfully measured in the range 25–1200 nm. In general, experiments with glass and metal-coated glass have yielded results in good agreement with the Lifshitz theory: the expected power law for retarded forces was obtained, and the values found for the magnitudes of the forces agreed with theory within a factor of two. For a review of the earlier work, see Israelachvili and Tabor (1973), Israelachvili and Ninham (1977); for more recent work, see Derjaguin et al. (1978), and van Blokland and Overbeek (1978, 1979).

Unfortunately, experiments with glass were unable to provide accurate results for separations less than about 10 nm. For this, a much smoother surface than could be obtained with glass was needed. This problem was resolved by making use of naturally occurring muscovite mica, which may be cleaved to provide molecularly smooth surfaces over large areas, and by employing a multiple beam interferometry technique for measuring surface separations to within ± 0.1 nm. In this way Tabor and Winterton (1969) and later Israelachvili and Tabor (1972) and Coakley and Tabor (1978) measured the van der Waals forces between curved mica surfaces or metal-coated surfaces in air in the range 2–130 nm, where the agreement with theory was generally within 30%. Experiments were also carried out down to separations of 1.4 nm with a surfactant monolayer of thickness 2.5 nm deposited on each mica surface. The results showed that for separations greater than about 5 nm the effective Hamaker constant is as for bulk mica, but that for separations less than 3 nm it is about 25% less. Thus, the interaction at separations smaller than the layers' thickness was dominated by the properties of the monolayers (see Table XV), a result which is in accord with theoretical expectations.

More recently, experimental techniques have been developed to directly measure the van der Waals forces between solid surfaces in liquids (Israelachvili and Adams, 1978; Derjaguin et al., 1978). Figure 34 showed results obtained for the van der Waals force between two mica surfaces in various electrolyte solutions in the distance range 2–15 nm, where again the agreement with theory is within 30%. However, in liquids, unlike in air or vacuum, other forces are also usually present, such as long-range electric double-layer forces and—at very small separations, below a few molecular diameters—solvation forces. The one major limitation of the Lifshitz theory is that it treats the intervening solvent medium as a structureless continuum and consequently does not encompass solvation effects. We have seen in Chapter 7 that at very small separations the (solvation) force is expected to oscillate with distance with a periodicity equal to the molecular diameter—quite unlike the monotonic force law of the continuum Lifshitz theory. The short-range solvation forces between surfaces are described in Chapter 13.

Chapter 12

Electrostatic Forces between Surfaces in Liquids

12.1. The charging of surfaces in liquids

Situations in which van der Waals forces alone determine the total interaction are restricted to a few simple systems, for example, to interactions in vacuum or to nonpolar wetting films on surfaces, both of which were discussed in Chapter 11. In more complex, and more interesting, systems long-range electrostatic forces are also involved, and the interplay between these two interactions has many important consequences.

As mentioned earlier the van der Waals force between similar particles in a medium is always attractive, so that if only van der Waals forces were operating, we might expect all dissolved particles to stick together (coagulate) immediately and precipitate out of solution as a mass of solid material. Our own bodies would be subject to the same fate if we remember that we are composed of about 75% water. Fortunately this does not happen, because particles suspended in water or any liquid of high dielectric constant are usually charged and can be prevented from coalescing by repulsive electrostatic forces. The charging of surfaces can come about in two ways:

(i) by the ionization or dissociation of surface groups (e.g., the dissociation of protons from surface carboxylic groups [$-COOH \rightarrow -COO^- + H^+$], which leaves behind a negatively charged surface) and

(ii) by the adsorption (binding) of ions from solution onto a previously uncharged surface (e.g., the binding of Ca^{2+} onto the zwitterionic head groups of lipid bilayer surfaces, which charges the surfaces positively).

The adsorption of cations from solution can, of course, also occur onto anionic surface sites (e.g., the adsorption of Ca^{2+} onto COO^- sites vacated by H^+ or Na^+). This is known as *ion exchange*. Whatever the *charging mechanism*, the final surface charge is balanced by an equal but oppositely charged *atmosphere of counterions* in rapid thermal motion close to the surface, known as the *diffuse double layer* (Fig. 37).

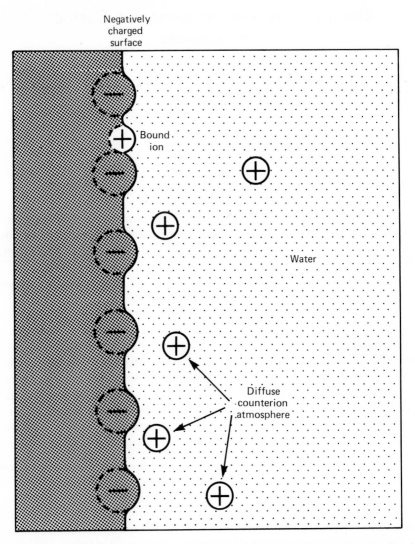

Fig. 37 Ions bound to a surface are not rigidly bound but can exchange with other ions in solution; their lifetime on a surface can be as short as 10^{-9} s or as long as many hours.

Two similarly charged surfaces always repel each other electrostatically in solution. Zwitterionic surfaces (i.e., those characterised by having surface dipoles but no net charge) also interact electrostatically with each other, though here we shall find that the force can be attractive or repulsive.

12.2. Charged surfaces in water (no added electrolyte)

In the following sections we shall consider the counterion distribution and force between two similarly charged planar surfaces in a *pure* liquid such as water, where (apart from H_3O^+ and OH^- ions) the only ions in the solution are those that have come off the surfaces. Such systems occur when, for example, colloidal particles, clay sheets, surfactant micelles, or bilayers whose surfaces contain ionizable groups interact in water and also when thick films of water build up (condense) on an ionizable surface such as glass. But first we must consider some fundamental equations that describe the counterion distribution between two charged surfaces in solution.

12.3. The Poisson–Boltzmann equation

For the case when only counterions are present in solution, the chemical potential of any ion may be written as

$$\mu = ze\psi + kT \log \rho, \tag{12.1}$$

where ψ is the electrostatic potential and ρ the number density of ions of valency z at any point x between two surfaces (Fig. 38). Since only differences in potential are ever physically significant, we may set $\psi_0 = 0$ at the midplane ($x = 0$), where $\rho = \rho_0$ and also $(d\psi/dx)_0 = 0$ by symmetry.

From the equilibrium requirement that the chemical potential be uniform throughout, Eq. (12.1) then gives us the expected Boltzmann distribution of counterions at any point x:

$$\rho = \rho_0 e^{-ze\psi/kT}. \tag{12.2}$$

One further important fundamental equation will be required. This is the well-known *Poisson equation*:

$$-\varepsilon\varepsilon_0(d^2\psi/dx^2) = \text{net charge density at } x = ze\rho, \tag{12.3}$$

which when combined with the Boltzmann distribution, Eq. (12.2), above gives the *Poisson–Boltzmann equation*:

$$d^2\psi/dx^2 = -(ze\rho_0/\varepsilon\varepsilon_0)e^{-ze\psi/kT}. \tag{12.4}$$

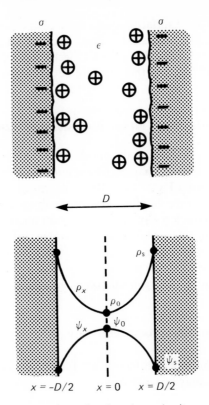

Fig. 38 Two negatively charged surfaces of surface charge density σ separated a distance D in water. The only ions in the interspace are the counterions that have dissociated from the surfaces. The counterion density profile ρ_x and electrostatic potential ψ_x are shown schematically in the lower part of the figure.

This is a (nonlinear) second-order differential equation, and to solve for ψ we need two *boundary conditions*, which determine the two integration constants. The first boundary condition follows from the symmetry requirement that the field must vanish at the midplane [i.e., $(d\psi/dx)_0 = 0$]. The second boundary condition follows from the requirement of electroneutrality [i.e., that the total charge of the counterions in the interspace must be equal (and opposite) to the charge on the surfaces]. If σ is the surface charge density on each surface (in C m^{-2}) and D is the distance between the surfaces, then

$$\sigma = -\int_0^{D/2} ze\rho \, dx = +\varepsilon\varepsilon_0 \int_0^{D/2} (d^2\psi/dx^2) \, dx$$

$$= \varepsilon\varepsilon_0 (d\psi/dx)_{D/2} = \varepsilon\varepsilon_0 (d\psi/dx)_{\text{surface}}. \tag{12.5}$$

Eq. (12.5) gives an important *boundary condition* relating σ to the electric field $\pm(d\psi/dx)_s$ at either surface (at $D/2$ and $-D/2$).

12.4. The contact value theorem

There exists an important general relation between the concentrations of counterions at either surface and at the midplane. Differentiating Eq. (12.2) and then using Eq. (12.4) we obtain

$$\frac{d\rho}{dx} = -\frac{ze\rho_0}{kT}e^{-ze\psi/kT}\left(\frac{d\psi}{dx}\right) = \frac{\varepsilon\varepsilon_0}{kT}\left(\frac{d\psi}{dx}\right)\left(\frac{d^2\psi}{dx^2}\right)$$

$$= \frac{\varepsilon\varepsilon_0}{2kT}\frac{d}{dx}\left(\frac{d\psi}{dx}\right)^2,\tag{12.6}$$

hence

$$\rho_x - \rho_0 = \int_0^x d\rho = \frac{\varepsilon\varepsilon_0}{2kT}\int_0^x d\left(\frac{d\psi}{dx}\right)^2 = +\frac{\varepsilon\varepsilon_0}{2kT}\left(\frac{d\psi}{dx}\right)_x^2$$

so that

$$\rho_x = \rho_0 + \frac{\varepsilon\varepsilon_0}{2kT}\left(\frac{d\psi}{dx}\right)_x^2,\tag{12.7}$$

which gives ρ at any point x in terms of ρ_0 at the midplane and $(d\psi/dx)^2$ at x. In particular at the surface, $x = D/2$, we find using Eq. (12.5) that

$$\rho_s = \rho_0 + \sigma^2/2\varepsilon\varepsilon_0 kT.\tag{12.8}$$

This is known as the *contact value theorem*. It is a relation (independent of valency z) giving the concentration of counterions at the surface in terms of the surface charge density σ and the counterion concentration at the midplane. Note that ρ_s never falls below $\sigma^2/2\varepsilon\varepsilon_0 kT$ even for isolated surfaces (i.e., for two surfaces far apart when $\rho_0 \to 0$). For example, consider an isolated surface in water of charge density $\sigma = 0.2\ \mathrm{C\ m^{-2}}$; this is a typical value for a fully ionized surface and corresponds to one elementary charge per 0.8 nm^2 or about 1.25×10^{18} charges per square meter. The concentration of counterions at the surface at 293 K is therefore

$$\rho_s = \sigma^2/2\varepsilon\varepsilon_0 kT$$

$$= (0.2)^2/2 \times 80 \times 8.85 \times 10^{-12} \times 1.38 \times 10^{-23} \times 293$$

$$= 7.0 \times 10^{27}\ \mathrm{m^{-3}},$$

which is about 12 M. If these surface counterions are considered to occupy a layer of thickness ~ 0.2 nm, then the above value for ρ_s corresponds to

$$7 \times 10^{27} \times 0.2 \times 10^{-9} \simeq 1.4 \times 10^{18} \quad \text{ions/m}^2,$$

which is about the same as the surface charge density σ. This is an interesting result, for it shows that regardless of the counterion distribution profile ρ_x away from a surface (Section 12.5), most of the counterions that effectively balance the surface charge are located in the first few angstroms from the surface (Jönsson *et al.*, 1980). However, for lower surface charge densities, since $\rho_s \propto \sigma^2$ the diffuse layer of counterions extends well beyond the surface.

12.5. Counterion concentration profile away from a surface

The above equations are quite general and are the starting point of all theoretical computations of the ionic distributions near planar surfaces, even when the solution contains added electrolyte (Sections 12.10–12.16). To proceed further for the specific case of counterions only (Fig. 38) we must now solve the Poisson–Boltzmann equation, Eq. (12.4), which can be satisfied by

$$\psi = (kT/ze)\log(\cos^2 Kx) \tag{12.9}$$

or

$$e^{-ze\psi/kT} = 1/\cos^2 Kx, \tag{12.10}$$

where K is a constant given by

$$K^2 = (ze)^2 \rho_0/2\varepsilon\varepsilon_0 kT. \tag{12.11}$$

With this form for the potential we see that $\psi = 0$ and $d\psi/dx = 0$ at $x = 0$ for all K, as required. To solve for K we differentiate Eq. (12.9) and then use Eq. (12.5) to obtain

$$d\psi/dx = -(2kTK/ze)\tan Kx,$$
$$(d\psi/dx)_s = -(2kTK/ze)\tan(KD/2) = +\sigma/\varepsilon\varepsilon_0. \tag{12.12}$$

The counterion distribution profile

$$\rho_x = \rho_0 e^{-ze\psi/kT} = \rho_0/\cos^2 Kx \tag{12.13}$$

is therefore known once K is determined from Eq. (12.12) in terms of σ and D.

For example, if the two surfaces with $\sigma = 0.2$ C m^{-2} are 2 nm apart ($D = 2$ nm), then from Eq. (12.12) we find for $z = 1$ that $K = 1.3361 \times 10^9$ m^{-1} at 293 K. From Eq. (12.11) this means that $\rho_0 = 0.4 \times 10^{27}$ m^{-3}, so that at the surface $\rho_s = \rho_0/\cos^2(KD/2) = 7.4 \times 10^{27}$ m^{-3}. The same result is also

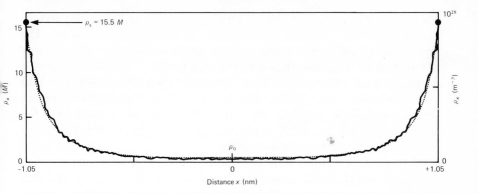

Fig. 39 Monovalent counterion concentration profile between two charged surfaces ($\sigma =$ 0.224 C m^{-2}, corresponding to one electronic charge per 0.714 nm^2) a distance 2.1 nm apart in water. The smooth curve is obtained from the Poisson–Boltzmann equation; the other is from a Monte Carlo simulation by Jönsson *et al.* (1980).

immediately obtainable from Eq. (12.8) since, as we have previously established, $\sigma^2/2\varepsilon\varepsilon_0 = 7.0 \times 10^{27}$ m^{-3}. Thus, the counterion concentration at each surface ρ_s is about 18.5 times greater than at the midplane ρ_0, which is only 1 nm away.

Figure 39 shows how the counterion concentration varies with distance for the case of $\sigma = 0.224$ C m^{-2}, $D = 2.1$ nm, as calculated on the basis of (i) the Poisson–Boltzmann equation, Eq. (12.13), and (ii) a Monte Carlo simulation of the same system. The agreement is quite good though the Monte Carlo results gives a slightly higher counterion concentration very near the surfaces compensated by a lower concentration in the central region between the two surfaces.

12.6. Origin of the ionic distribution, surface potential, and pressure

Before we proceed to calculate the pressure between two planar surfaces it is instructive to discuss, in qualitative terms, how the counterion distribution, potential, and pressure between two surfaces arise. The first thing to notice is that if there were no ions between two charged surfaces, there would be no electric field in the region between them. This is because the field emanating from a planar charged surface, $E = \sigma/2\varepsilon\varepsilon_0$, is uniform away from the surface. Thus, the two opposing fields emanating from two plane parallel surfaces cancel out to zero between them. When the counterions are introduced into the intervening region they do *not* experience an attractive electrostatic force towards each surface. The reason why the counterions build up at each surface is simply because of their mutual repulsion and is similar to the accumulation of mobile charges on the surface of any charged conductor. The repulsive electrostatic interaction between the counterions and their entropy of mixing *alone* determine their concentration profile ρ_x, the potential profile ψ_x, and ultimately the repulsive force experienced by the two surfaces (Jönsson *et al.*, 1980); and we may further note that in all the theoretical derivations so far encountered the only way the surface charge density σ enters into the picture is through Eq. (12.5), which is simply a statement about the total number of counterions in the interspace.

Further, if the locations of the two ionizable surfaces were not at $x = \pm D/2$ but farther away from the boundary of the aqueous medium (Fig. 40), the ionic distribution ρ_x, potential ψ_x, and pressure in the medium would not change, but the potential would be different if it were measured at $x = \pm(D/2 + \delta)$. This is the origin of the so called *Stern layer* effect (Verwey and Overbeek, 1948; Hiemenz, 1977). The Stern layer is that narrow space separating the charged plane from the steric wall from which the ionic atmosphere begins to obey the Poisson–Boltzmann equation. The thickness of the Stern layer δ is of the order of one or two angstroms and reflects the finite size of surface ionizable groups and counterions, as illustrated in Fig. 40.

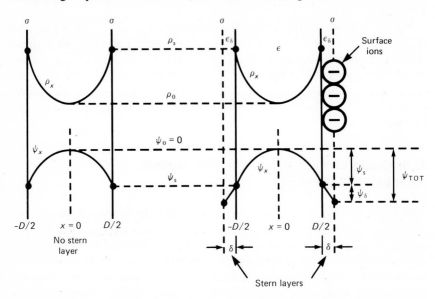

Fig. 40 Stern layers of thickness δ at each surface dividing the planes of fixed charge density σ from the boundary surfaces of the aqueous solution. The counterion density and electrostatic potential within the aqueous region between $x = D/2$ and $x = -D/2$, and the pressure between the two surfaces, are independent of δ. But there is an additional linear drop in potential across the Stern layer given by $\psi_\delta = \delta\sigma/\varepsilon_\delta\varepsilon_0$.

12.7. The pressure between two charged surfaces in water

The repulsive pressure P of the counterions at any position x is related to the chemical potential μ via Eq. (2.9) and may be written as

$$(\partial P/\partial x)_T = \rho(\partial \mu/\partial x)_T. \tag{12.14}$$

Thus, at constant temperature

$$P = -\int_x^\infty \{ze\rho(\mathrm{d}\psi/\mathrm{d}x)\,\mathrm{d}x + kT(\mathrm{d}\rho/\mathrm{d}x)\,\mathrm{d}x\}. \tag{12.15}$$

Replacing $ze\rho$ by the Poisson equation and using the relation

$$\frac{\mathrm{d}}{\mathrm{d}x}\left(\frac{\mathrm{d}\psi}{\mathrm{d}x}\right)^2 = 2\left(\frac{\mathrm{d}\psi}{\mathrm{d}x}\right)\left(\frac{\mathrm{d}^2\psi}{\mathrm{d}x^2}\right),$$

this becomes

$$P = -\tfrac{1}{2}\varepsilon\varepsilon_0(\mathrm{d}\psi/\mathrm{d}x)_x^2 + kT(\rho_x - \rho_\infty), \tag{12.16}$$

where in the present case $\rho_\infty = 0$ since there are no electrolyte ions in the bulk solution. The above important equation gives the pressure at any point x between two surfaces. At equilibrium this pressure must be uniform throughout (i.e., independent of x) and is also the pressure acting on the two surfaces. To verify this we note that using Eq. (12.7) the above may be written as

$$P = kT(\rho_0 - \rho_x) + kT(\rho_x - \rho_\infty) = kT(\rho_0 - \rho_\infty) \tag{12.17a}$$

$$= kT\rho_0 \quad \text{when} \quad \rho_\infty = 0, \tag{12.17b}$$

which is independent of x and only depends on ρ_0 (i.e., on σ and D). It is instructive to insert Eq. (12.8) into the above, whence we obtain

$$P = kT\rho_0 = kT[\rho_s - \sigma^2/2\varepsilon\varepsilon_0 kT],$$

that is,

$$P(D) = kT[\rho_s(D) - \rho_s(\infty)]. \tag{12.18}$$

Thus, the pressure is directly proportional to the increase in the ion concentration at the surfaces as they approach each other. This important result, a more general form of the *contact value theorem*, is always valid so long as the surface charge remains independent of D. It also applies when electrolyte ions i are present in the solution (where ρ_s becomes replaced by $\Sigma_i \, \rho_{si}$) and also to other types of interactions [e.g., solvation interactions, where ρ_s is now the surface concentration of solvent molecules (Chapter 13)].

Returning to Eq. (12.17b) the pressure may also be expressed in terms of K, as given by Eq. (12.11), by

$$P = kT\rho_0 = 2\varepsilon\varepsilon_0(kT/ze)^2 K^2. \tag{12.19}$$

As an example let us apply this result to the case of two surfaces with $\sigma = 0.2$ C m^{-2} at $D = 2$ nm apart, for which we previously found $K = 1.336 \times 10^9$ m^{-1}. The repulsive pressure between them is therefore 1.7×10^6 N m^{-2}, or about 17 atm. This repulsion exceeds by far any possible van der Waals attraction at this separation; for a typical Hamaker constant of $A \approx 10^{-20}$ J the van der Waals attractive pressure would be only $A/12\pi D^3 \approx 3 \times 10^4$ N m^{-2} or about 0.3 atm.

The above equations have been used successfully to account for the equilibrium spacings of ionic surfactant and lipid bilayers in water (Cowley et al., 1978). Figure 41 shows experimental results obtained for the repulsive pressure between charged bilayer surfaces in water (cf. Fig. 28h), together with the theoretical curve based on Eq. (12.19). The agreement is very good down to $D \approx 2$ nm and shows that the effective charge density of the anionic lipid headgroups is about $1e$ per 14 nm^2. At smaller distances the measured forces are more repulsive than expected due to solvation or "hydration" forces, arising from the additional energy needed to dehydrate these very hydrophilic head groups (Chapter 13).

Repulsive electrostatic forces are also believed to control the long-range swelling of clays in water. Most naturally occurring clays are composed of lamellar aluminosilicate sheets about 1 nm thick whose surfaces dissociate in water giving off Na$^+$, K$^+$, and Ca^{2+} ions, and when placed in water they can swell to more than 10 times their original volume (Norrish, 1954). The swelling of clays is, however, a complex matter and also involves hydration effects at surface separations below about 3 nm (van Olphen, 1977).

In the case of charged spherical particles (e.g., latex particles) in water the long-range electrostatic repulsion between them can result in an ordered lattice of particles even when the distance between them is well in excess of their diameter (Takano and Hachisu, 1978). In such systems (cf. Fig. 28g) colloidal particles attempt to get as far apart from each other as possible but, being constrained in a finite solution, are forced to arrange themselves into an ordered lattice (for a review see Forsyth et al., 1978).

Parsegian (1966) and Jönsson and Wennerström (1981) extended the theoretical analysis to the interactions of both cylindrical and spherical structures in water, and the results were used to analyse the relative stability of ionic surfactant aggregates in water. Such "micellar" aggregates form spontaneously in water, and there is a general tendency for them to change from spherical to cylindrical to lamellar (bilayer) structures as the amount of water is reduced (see also Part III). Wennerström et al. (1982) also compared

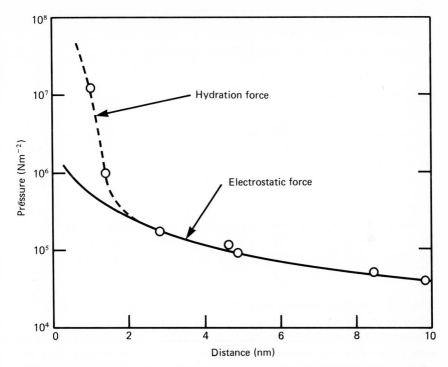

Fig. 41 Measured repulsive pressure between charged bilayer surfaces in water. The bilayers were composed of 90% lecithin, a neutral zwitterionic lipid, and 10% phosphatidylglycerol, a negatively charged lipid. For full ionization, the surface charge density should be one electronic charge per 7 nm^2, whereas the theoretical line through the experimental points suggests one charge per 14 nm^2 (i.e., about 50% ionization). Below 2 nm there is an additional repulsion due to "hydration forces." [From Cowley *et al.* (1978). © 1978 American Chemical Society.]

the ionic density distributions and interaction energies and pressures between planar surfaces, spheres, and cylinders with Monte Carlo simulations and found that for monovalent counterions, the Monte Carlo results always yield pressures 10–50% lower than those obtained from the Poisson–Boltzmann equation (cf. the lower value for ρ_0 obtained from the Monte Carlo simulation in Fig. 39).

Other factors that must be important in determining the ionic distributions and pressures in real systems involve

(i) the finite sizes of ions (both the counterions and those bound to surfaces),

(ii) the fact that the surface charges are discrete and not "smeared out" as has been assumed,

(iii) image forces, and

(iv)   perhaps of greatest importance, the short-range ion–solvent and solvent–solvent interactions (solvation or hydration forces) where the solvent no longer behaves as a continuum (Carnie and Chan, 1982).

12.8. Thick wetting films

At large distances D and high surface charge densities σ the value of $(KD/2)$ in Eq. (12.12) must approach $\pi/2$ (i.e., $K \to \pi/D$). In this limit the pressure, Eq. (12.19), therefore becomes

$$P(D) = 2\varepsilon\varepsilon_0(\pi kT/ze)^2/D^2, \tag{12.20}$$

which is known as the *Langmuir equation*. The Langmuir equation has been used to account for the equilibrium thickness of thick wetting films of water on glass surfaces (Figs. 28f and 42). Here the water–air surface replaces the midplane of Fig. 38 so that for a film of thickness $d = D/2$, we have

$$P(d) = \varepsilon\varepsilon_0(\pi kT/ze)^2/2d^2, \tag{12.21}$$

which is sometimes referred to as the *disjoining pressure* of the film. This repulsive pressure is entirely analogous to the repulsive van der Waals force across liquid helium films that causes liquid helium to climb up the walls of containers. This was discussed in Section 11.5, where we saw that the equilibrium thickness d of a film at a height H above the surface of the liquid is given by Eq. (11.18) as $P(d)v = -mgH$, where v and m are the molecular

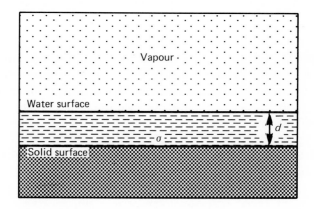

Fig. 42 A water film on a charged (ionizable) glass surface will tend to thicken because of the repulsive "disjoining pressure" of the counterions in the film. If the vapour over the film is saturated, the film will grow indefinitely, but if it is unsaturated, the equilibrium thickness d will be finite as given by Eq. (12.22).

volume and mass. Note that at the height H, the vapour pressure of the liquid vapour is given by the Boltzmann factor

$$\rho = \rho_{sat}e^{-mgH/kT},$$

where ρ_{sat} is the saturated vapour pressure (at $H = 0$). Thus, if water condenses on a surface from undersaturated vapour, its equilibrium thickness d will be determined by

$$P(d) = -mgH/v = (kT/v)\log(\rho/\rho_{sat}) \qquad (12.22)$$

(i.e., the film thickness will increase to infinity as ρ approaches ρ_{sat}).

Returning to the Langmuir equation, Langmuir (1938) first applied Eq. (12.21) to explain why the rise of water up a capillary tube is higher than expected: since the water also wets the inner surface of the capillary the effective radius of the capillary is smaller than the dry radius and this leads to a higher capillary rise than expected. Derjaguin and Kusakov (1939) measured how the thickness of a water film on a quartz glass surface decreased when an air bubble was progressively pressed down on the film. The results were in rough agreement with the Langmuir equation. Read and Kitchener (1969) repeated these measurements and again found only rough agreement between theory and experiment: the measured film thicknesses (in the range 30–130 nm) were consistently thicker (by 10–20 nm) than expected theoretically, an effect that could be accounted for if the water–air interface is negatively charged so that for a given pressure P the film thickness would be higher than given by Eq. (12.21), which assumes $\sigma = 0$ and $d\psi/dx = 0$ at that interface. However, Derjaguin and Churaev (1974), and Pashley and Kitchener (1979) used the vapour pressure control method to measure the equilibrium film thickness and found that for $d < 30$ nm the films are *much* thicker than expected from Eq. (12.22), an effect that has been attributed to either repulsive hydration forces or to the presence of small amounts of soluble contaminants in the films (Pashley, 1980).

12.9. Charge regulation

At small separations, as $D \to 0$, it is easy to verify that $K^2 \to -\sigma ze/\varepsilon\varepsilon_0 kTD$ (note that σ and z must have opposite signs). Thus, the repulsive pressure P of Eq. (12.19) approaches infinity according to $P = -2\sigma kT/zeD$, and the surface concentration of ions ρ_s also rises dramatically as D approaches zero. (Since $-2\sigma/ze$ is the number of counterions per unit area in the interspace this limiting pressure is simply the osmotic pressure of an ideal gas at the same density as the trapped counterions.) An infinite pressure is, of course, unrealistic and arises from the assumption that σ = constant (i.e., that the surfaces remain ionized even when there is a very large pressure pushing the ions against the surfaces). In practice when two surfaces are finally forced into molecular contact the counterions are forced to readsorb onto their original surface sites. Thus, as D approaches zero the surface charge density σ also falls (i.e., σ becomes a function of D). This is known as *charge regulation*. The effect of charge regulation is always to reduce the effective electrostatic repulsion below that calculated on the assumption of constant surface charge and will be discussed again in Section 12.16.

12.10. Charged surfaces in electrolyte solutions

It is far more common for charged surfaces or particles to interact across a solution that already contains electrolyte ions (i.e., dissociated inorganic salts). In animal fluids, ions are present in concentrations of about 0.2 M, mainly NaCl or KCl with smaller amounts of $MgCl_2$ and $CaCl_2$. The oceans have a similar relative composition of these salts but of a higher total concentration, about 0.6 M. Note that even "pure water" at pH 7 is strictly an electrolyte solution containing 10^{-7} M of H_3O^+ and OH^- ions, which can not always be ignored.

The existence of a bulk "reservoir" of electrolyte ions has a profound effect on the electrostatic potential and force between charged surfaces, and in the rest of this chapter we shall consider this interaction and the nature of the total interaction potential when the attractive van der Waals force is added. But to understand the electrostatic interaction of two surfaces it is necessary to first understand the ionic distribution adjacent to an isolated surface in contact with an electrolyte solution. Consider an isolated surface (or two surfaces far apart) in an aqueous electrolyte (Fig. 43). For convenience, we shall put $x = 0$ at the surface rather than at the midplane. Now, all the fundamental equations derived in the previous sections are applicable to solutions containing different types of ions i (of valency $\pm z_i$) so long as this is taken into account by expressing the net charge density at any point x as $\Sigma_i z_i e \rho_{xi}$ and the total ionic concentration (number density) as $\Sigma_i \rho_{xi}$.

Equation (12.2) for the Boltzmann distribution of ions i at x now becomes

$$\rho_{xi} = \rho_{\infty i} e^{-z_i e \psi_x / kT}, \qquad (12.23)$$

while at the surface, at $x = 0$,

$$\rho_{0i} = \rho_{\infty i} e^{-z_i e \psi_0 / kT}, \qquad (12.24)$$

where $\rho_{\infty i}$ is the ionic concentration of ions i in the bulk (at $x = \infty$) where

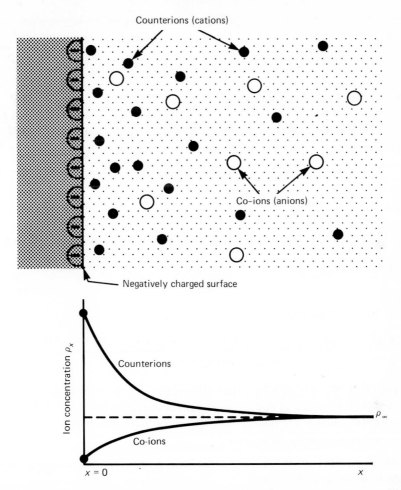

Fig. 43 Near a charged surface there is an accummulation of counterions (ions of opposite charge to the surface charge) and a depletion of co-ions, shown graphically below for a 1:1 electrolyte, where ρ_∞ is the bulk electrolyte concentration.

$\psi_\infty = 0$. For example, if we have a solution containing $H^+OH^- + Na^+Cl^- + Ca^{2+}Cl_2^-$, etc., we may write

$$[H^+]_x = [H^+]_\infty e^{-e\psi_x/kT}, \qquad [H^+]_0 = [H^+]_\infty e^{-e\psi_0/kT},$$

$$[Na^+]_x = [Na^+]_\infty e^{-e\psi_x/kT}, \qquad [Na^+]_0 = [Na^+]_\infty e^{-e\psi_0/kT},$$

$$[Ca^{2+}]_x = [Ca^{2+}]_\infty e^{-2e\psi_x/kT}, \qquad [Ca^{2+}]_0 = [Ca^{2+}]_\infty e^{-2e\psi_0/kT}, \qquad (12.25)$$

$$[Cl^-]_x = [Cl^-]_\infty e^{+e\psi_x/kT}, \qquad [Cl^-]_0 = [Cl^-]_\infty e^{+e\psi_0/kT},$$

where $[Na^+]$, etc., are expressed in some convenient concentration unit such as M ($1\ M = 1$ mol dm^{-3} and corresponds to a number density of $\rho = 6.022 \times 10^{26}$ m^{-3}).

12.11. The Grahame equation

Let us now find the total concentration of ions at an isolated surface of charge density σ. From Eq. (12.8) this is immediately given by

$$\sum_i \rho_{0i} = \sum_i \rho_{\infty i} + \sigma^2/2\varepsilon\varepsilon_0 kT \quad \text{(in number per m}^3). \tag{12.26}$$

Thus, for $\sigma = 0.2$ C m^{-2} (corresponding to one electronic charge per 0.8 nm^2 or 80 Å^2) at 25°C, we find $\sigma^2/2\varepsilon\varepsilon_0 kT = 7.0 \times 10^{27}$ m$^{-3} = 11.64\ M$. Thus, for a 1:1 electrolyte such as NaCl, the surface concentration of ions in this case is

$$[Na^+]_0 + [Cl^-]_0 = 11.64 + [Na^+]_\infty + [Cl^-]_\infty = 11.64 + 2[Na^+]_\infty$$
$$= 11.64 + 2[NaCl], \tag{12.27}$$

while for a 2:1 electrolyte such as CaCl$_2$,

$$[Ca^{2+}]_0 + [Cl^-]_0 = 11.64 + [Ca^{2+}]_\infty + [Cl^-]_\infty = 11.64 + 3[Ca^{2+}]_\infty$$
$$= 11.64 + 3[CaCl_2],$$

where $[NaCl]$ and $[CaCl_2]$ are the bulk molar concentrations of the salts. The ions at the surface are, of course, mainly the counterions (e.g., Na$^+$ or Ca^{2+} for a negatively charged surface), and their *excess* concentration at the surface over that in the bulk is seen to be

(i) dependent solely on the surface charge density σ (i.e., *independent* of their bulk concentration) and

(ii) of magnitude sufficient to balance most of the surface charge (cf. Sections 12.4 and 12.15).

We may now find the relation between the surface charge density σ and the surface potential ψ_0. Incorporating Eq. (12.24) into Eq. (12.26) we obtain for the case of a mixed NaCl + CaCl$_2$ electrolyte:

$$\sigma^2 = 2\varepsilon\varepsilon_0 kT \left(\sum_i \rho_{0i} - \sum_i \rho_{\infty i} \right)$$
$$= 2\varepsilon\varepsilon_0 kT \{ [Na^+]_\infty e^{-e\psi_0/kT} + [Ca^{2+}]_\infty e^{-2e\psi_0/kT}$$
$$+ [Cl^-]_\infty e^{+e\psi_0/kT} - [Na^+]_\infty - [Ca^{2+}]_\infty - [Cl^-]_\infty \}.$$

On further noting that $[Cl^-]_\infty = [Na^+]_\infty + 2[Ca^{2+}]_\infty$, the above becomes

$$\sigma^2 = 2\varepsilon\varepsilon_0 kT\{[Na^+]_\infty (e^{-e\psi_0/kT} + e^{+e\psi_0/kT} - 2)$$
$$+ [Ca^{2+}]_\infty (e^{-2e\psi_0/kT} + 2e^{e\psi_0/kT} - 3)\},$$

so that finally

$$\sigma = \sqrt{8\varepsilon\varepsilon_0 kT} \sinh(e\psi_0/2kT)\{[Na^+]_\infty + [Ca^{2+}]_\infty (2 + e^{-e\psi_0/kT})\}^{1/2}$$
$$= 0.117 \sinh(\psi_0/51.4)\{[NaCl] + [CaCl_2](2 + e^{-\psi_0/25.7})\}^{1/2} \qquad (12.28)$$

at 25°C, where $[NaCl] = [Na^+]_\infty$ and $[CaCl_2] = [Ca^{2+}]_\infty$ are in M, ψ_0 in mV, and σ in C m^{-2} (1 C m^{-2} corresponds to one electronic charge per 0.16 nm^2 or 16 Å^2). We shall now consider the implications of this important equation, known as the *Grahame equation*.

12.12. Surface charge and potential in the presence of monovalent ions

For an aqueous 1:1 electrolyte solution such as NaCl against a negatively charged surface of $\sigma = -0.2$ C m^{-2}, we obtain the potentials shown in the middle column of Table XVIII. Note that for no electrolyte, we obtain an infinite potential, which is unrealistic: a pure liquid such as water will always contain *some* dissociated ions. It is for this reason that we did not consider an

TABLE XVIII

Variation of surface potential with aqueous electrolyte concentration for a planar surface of charge density -0.2 C m^{-2} as deduced from the Grahame equation, Eq. (12.28).

1:1 Electrolyte Concentration (M)	ψ_0 (mV)	
	Pure 1:1 Electrolyte	Bulk Solution Also Contains 3×10^{-3} M 2:1 Electrolyte
0 (hypothetical)	$-\infty$	-106
10^{-7} (pure water)	-477	-106
10^{-4}	-300	-106
10^{-3}	-241	-106
10^{-2}	-181	-105
10^{-1}	-123	-100
1	-67	-66

isolated surface in the absence of bulk electrolyte ions in Section 12.5. From Table XVIII we find that at constant surface charge density the surface potential falls progressively as the electrolyte concentration rises. Incidentally, from the tabulated values of ψ_0 we can determine the ionic concentrations at the surface using Eq. (12.25). For example, in 10^{-7} M 1:1 electrolyte, where $\psi_0 \approx -477.1$ mV, we obtain $10^{-7} \times e^{477.1/25.69} = 11.64$ M for the counterions, and $10^{-7} \times e^{-477.1/25.69} \approx 10^{-15}$ M for the coions. In 1 M, where $\psi_0 = -67.0$ mV, we obtain 13.57 M and 0.07 M for the counterions and coions, respectively. As expected, the total concentrations at the surface agree exactly with those previously obtained using Eqs. (12.26) and (12.27).

In most cases ionizable surface sites are not fully dissociated but are partially neutralized by the binding of specific ions in the solution. Such ions are often referred to as *exchangeable ions*, in contrast to those *inert* ions that do not bind to the surface. For example, if only protons can bind to a negatively charged surface, the equilibrium condition at the surface is given by the familiar *mass action equation* (Payens, 1955). Thus, for the reaction

$$S^- + H^+ \rightleftharpoons SH \quad \text{at the surface,} \qquad (12.29)$$

we may express the proton concentration at the surface as $[H^+]_0$, the concentration or surface density of negative (dissociated) surface sites as $[S^-]_0$, and the density of undissociated sites as $[SH]_0$. $[S^-]_0$ is related to σ via $\sigma = -e[S^-]_0$. The surface *dissociation constant* K for the above "reaction" is defined by

$$K = \frac{[S^-]_0[H^+]_0}{[SH]_0} \qquad (12.30a)$$

$$= \frac{\sigma_0\alpha[H^+]_0}{\sigma_0(1-\alpha)} = \frac{\alpha}{(1-\alpha)}[H^+]_0 = \frac{\alpha}{(1-\alpha)}[H^+]_\infty e^{-e\psi_0/kT}, \qquad (12.30b)$$

where σ_0 is the maximum possible charge density (i.e., if all the sites are dissociated) and α is the fraction of sites actually dissociated. Thus, if half the sites are dissociated at $[H^+]_0 = 10^{-4}$ M, we have $K = [H^+]_0 = 10^{-4}$ M, which is often quoted in pK units (p$K = -\log_{10}[H^+]_0 = 4.0$ in this case).

For a mixed 1:1 electrolyte of NaCl + HCl, Eq. (12.30) can be combined with the Grahame equation to give

$$\sigma = \sigma_0\alpha = \sigma_0 K/(K + [HCl]e^{-\psi_0/25.7}),$$

$$= 0.117\sinh(\psi_0/51.4)\sqrt{[NaCl] + [HCl]}, \qquad (12.31)$$

in which both σ and ψ_0 can now be totally determined in terms of the maximum charge density σ_0 and dissociation constant K (assuming that there is no binding of Na^+ ions). It is clear from the above that if K is very large (weak binding of protons), $\sigma \approx \sigma_0 \approx$ constant, and we obtain the earlier result for the case of fixed surface charge density. However, if K takes on a more typical value, the effect can be quite dramatic. For example, if $K = 10^{-4}$ M, then for a surface of $\sigma_0 = -0.2$ C m^{-2} in a 0.1 M NaCl bulk solution at pH 7 we find $\psi_0 = -118$ mV and $\alpha = 0.91$ (not very different from Table XVIII); but at pH 5 we obtain $\psi_0 = -73$ mV and $\alpha = 0.36$ (i.e., only 36% of the sites remain dissociated even though the bulk concentration of HCl is a mere 0.01% of the NaCl concentration). (Under such conditions the proton is referred to as a *potential determining ion*.) Thus, both ψ_0 *and* σ will vary as the salt concentration or pH is changed, but the surface will always remain negatively charged. More generally, a surface may contain both anionic (e.g., acidic) and cationic (e.g., basic) groups to which various cations and anions can bind. Such surfaces are known as *amphoteric surfaces*, and the *competitive adsorption* of ions to them can be analysed by assigning a binding constant to each ion type, similar to Eq. (12.30a), and then incorporating these into the Grahame equation (Healy and White, 1978). The charge density of amphoteric surfaces (e.g., protein surfaces) can be negative or positive depending on the electrolyte conditions. At the *isoelectric point* there are as many negative charges as positive charges on the surface so that the mean charge density is zero ($\sigma = 0$), though it is well to remember that there may still be patches of high local charge density.

12.13. Effect of divalent ions

The presence of divalent cations has a dramatic effect on the surface potential and counterion distribution at a negatively charged surface. For example, if all the NaCl solutions of Table XVIII also contain 3×10^{-3} M CaCl$_2$, the Grahame equation gives the potentials shown in the last column of Table XVIII. We see that even at constant surface charge density relatively small amounts of divalent ions substantially lower the magnitude of ψ_0 (in fact, about 100 times more effectively than increasing the concentration of monovalent salt). Indeed, ψ_0 is determined solely by the divalent cations once their concentration is greater than about 3% of the monovalent ion concentration; and for 2:1 electrolyte concentrations above a few mM, typical surface potentials are usually much smaller than -100 mV irrespective of the 1:1 electrolyte concentration. Further, even when the bulk concentration of Ca^{2+} is much smaller than that of Na^+ the surface may still have a much higher concentration of Ca^{2+}. For example, in 100 mM NaCl + 3 mM CaCl$_2$ where $\psi_0 = -100$ mV (Table XVIII) the concentration of Ca^{2+} at the surface is $[Ca^{2+}]_0 \approx 3 \times 10^{-3} e^{200/25.7} \approx 7$ M compared to $[Na^+]_0 \approx$ $0.1 e^{100/25.7} \approx 5$ M. At such high surface concentrations (of *doubly* charged ions) divalent ions often bind chemically to negative surface sites, thereby lowering σ and reducing ψ_0 even further; and it is not unusual for surfaces to be completely neutralized ($\sigma \to 0, \psi_0 \to 0$) in the presence of only mM amounts of Ca^{2+} or even smaller amounts of trivalent ions such as La^{3+}.

As in the case of monovalent ion binding, the effect of divalent ion binding can be dealt with quantitatively by incorporating the appropriate binding constant into the Grahame equation (Healy and White, 1978; McLaughlin *et al.*, 1981), and when many different ionic species (e.g., Ca^{2+}, H^+) compete for binding sites the variation of ψ_0 and σ with electrolyte concentration and pH can be quite complex. In most cases ion binding tends to lower both σ and ψ_0 as the concentrations of these ions increase, and we may anticipate that such effects lead to a substantial reduction in the repulsive electrostatic force between surfaces.

12.14. The electric double layer and the Debye length

For low potentials, below about 25 mV, the Grahame equation simplifies to

$$\sigma = \varepsilon \varepsilon_0 \kappa \psi_0, \qquad (12.32)$$

where

$$\kappa = \left(e^2 \sum_i \rho_{\infty i} z_i^2 / \varepsilon \varepsilon_0 kT \right)^{1/2}. \qquad (12.33)$$

Thus, the potential becomes proportional to the surface charge density. Equation (12.32) is the same as that of a capacitor whose two plates, separated by a distance $1/\kappa$, have charge densities $\pm \sigma$ and potential difference ψ_0. This analogy with a charged capacitor has given rise to the name *diffuse electric double layer* for describing the ionic atmosphere near a charged surface, whose characteristic length or "thickness" is the *Debye length*, $1/\kappa$.

At 25°C the magnitude of the Debye length is

$$1/\kappa \begin{cases} = 0.304/\sqrt{[NaCl]} & \text{nm for } 1:1 \text{ electrolytes (e.g., NaCl),} & (12.34) \\ = 0.176/\sqrt{[CaCl_2]} & \text{nm for } 2:1 \text{ electrolytes (e.g., } CaCl_2), & (12.35) \\ = 0.152/\sqrt{[MgSO_4]} & \text{nm for } 2:2 \text{ electrolytes (e.g., } MgSO_4). & (12.36) \end{cases}$$

For example, for NaCl solution, $1/\kappa = 30.4$ nm at 10^{-4} M, 0.96 nm at 0.1 M, and 0.3 nm at 1 M.

12.15. Variation of potential and ionic concentrations away from a charged surface

The potential gradient at any distance x from an isolated surface is given by Eq. (12.7):

$$\sum_i \rho_{xi} = \sum_i \rho_{\infty i} + \frac{\varepsilon\varepsilon_0}{2kT}\left(\frac{d\psi}{dx}\right)^2. \tag{12.37}$$

For a 1:1 electrolyte this gives

$$d\psi/dx = \sqrt{8kT/\varepsilon\varepsilon_0}\,\sinh(e\psi_x/2kT),$$

which may be readily integrated using the integral $\int \operatorname{csch} X\, dX = \log\tanh(X/2)$ to yield

$$\psi_x = \frac{2kT}{e}\log\left[\frac{1 + \gamma e^{-\kappa x}}{1 - \gamma e^{-\kappa x}}\right] \approx \frac{4kT}{e}\gamma e^{-\kappa x}, \tag{12.38}$$

where

$$\gamma = \tanh(e\psi_0/4kT) \approx \tanh[\psi_0(\mathrm{mV})/103]. \tag{12.39}$$

This is known as the *Gouy–Chapman theory*. For low potentials ψ_0, Eq. (12.38) reduces to the so-called *Debye–Hückel approximation*:

$$\psi_x \approx \psi_0 e^{-\kappa x}, \tag{12.40}$$

where again the Debye length $1/\kappa$ appears as the characteristic decay length of the potential [see Verwey and Overbeek (1948) and Hiemenz (1977) for a fuller discussion of the Gouy–Chapman and Debye–Hückel theories].

For a pure 2:1 electrolyte, a similar calculation (Grahame, 1953) gives the following relation between ψ_x and ψ_0 (for $\psi_0 > 0$):

$$\left(\frac{\sqrt{1 + 2e^{e\psi_x/kT}} - \sqrt{3}}{\sqrt{1 + 2e^{e\psi_x/kT}} + \sqrt{3}}\right) = \left(\frac{\sqrt{1 + 2e^{e\psi_0/kT}} - \sqrt{3}}{\sqrt{1 + 2e^{e\psi_0/kT}} + \sqrt{3}}\right)e^{-\kappa x}, \tag{12.41}$$

which for low ψ_0 again reduces to $\psi_x = \psi_0 e^{-\kappa x}$.

We now have all the equations needed for computing the ionic distributions away from a charged surface. For a 1:1 electrolyte, this is given by inserting Eq. (12.38) into Eq. (12.23) or (12.25). Figure 44 shows the variation of ψ_x and ρ_x for a 0.1 M 1:1 electrolyte, together with a Monte Carlo simulation for comparison. Note how the counterion density approaches the bulk value much faster than would be indicated by the Debye length. Indeed, for such a high surface charge density and ψ_0 the counterion distribution very near the surface is largely independent of electrolyte concentration, and it is left as an exercise for the reader to verify that even in 10^{-4} M the counterion profile over the first few angstroms is not much different from that in 0.1 M (so long as σ remains the same).

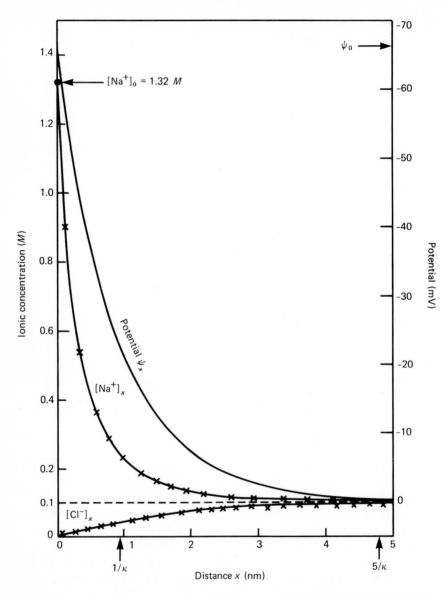

Fig. 44 Potential and ionic density profiles for a 0.1 M monovalent electrolyte such as NaCl near a surface of charge density $\sigma = -0.0621$ C m^{-2} (about one electronic charge per 2.6 nm^2), calculated from Eqs. (12.38) and (12.25) with $\psi_0 = -66.2$ mV obtained from the Grahame equation. The crosses are the Monte Carlo results of Torrie and Valleau (1979).

12.16. The electrostatic double-layer interaction between charged surfaces in electrolyte

The interaction pressure between two identically charged surfaces in an electrolyte solution (Fig. 45) can be derived quite simply as follows. First, we note that at any point x the pressure is given by Eq. (12.16) as

$$P = -\frac{1}{2}\varepsilon\varepsilon_0\left(\frac{d\psi}{dx}\right)_x^2 + kT\left(\sum_i \rho_{xi} - \sum_i \rho_{\infty i}\right). \tag{12.42}$$

Second, from Eq. (12.7) we have

$$\sum_i \rho_{xi} = \sum_i \rho_{mi} + \frac{\varepsilon\varepsilon_0}{2kT}\left(\frac{d\psi}{dx}\right)_x^2, \tag{12.43}$$

where $\sum_i \rho_{mi}$ is the total ionic concentration at the midplane, at $x = \frac{1}{2}D$. Incorporating Eq. (12.43) into Eq. (12.42) yields

$$P = kT\left(\sum_i \rho_{mi} - \sum_i \rho_{\infty i}\right), \tag{12.44}$$

which is essentially the same as Eq. (12.17) and shows that P is simply the excess osmotic pressure of the ions in the midplane over the bulk pressure. Since $\sum_i \rho_{\infty i}$ is known from the bulk electrolyte concentration the problem reduces to finding the midplane concentration of ions, and it is here that certain assumptions have to be made to obtain an analytic result (Verwey and Overbeek, 1948). For a 1:1 electrolyte such as NaCl, Eq. (12.44) may be written as

$$P = kT\rho_\infty[(e^{-e\psi_m/kT} - 1) + (e^{+e\psi_m/kT} - 1)] \approx e^2\psi_m^2\rho_\infty/kT, \tag{12.45}$$

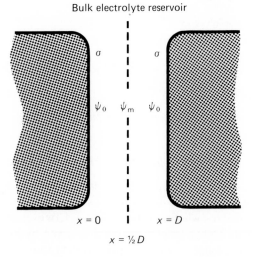

Bulk electrolyte reservoir

σ σ

ψ_0 ψ_m ψ_0

$x = 0$ $x = D$

$x = \frac{1}{2}D$

Fig. 45

which assumes that the midplane potential ψ_m (not the surface potential) is small. If we further assume that ψ_m is simply the sum of the potentials of each surface at $x = \frac{1}{2}D$ as previously derived for an isolated surface, then Eq. (12.38) gives $\psi_m \approx (8kT\gamma/e)e^{-\kappa D/2}$. Inserting this into Eq. (12.45) gives the final result for the repulsive pressure between two planar surfaces:

$$P = 64kT\rho_\infty \gamma^2 e^{-\kappa D} = (1.59 \times 10^8)[\text{NaCl}]\gamma^2 e^{-\kappa D} \quad \text{N m}^{-2}, \quad (12.46)$$

where we note that $\gamma = \tanh(e\psi_0/4kT)$ can never exceed unity. The above equation is known as the *weak overlap approximation*.

The interaction free energy per unit area corresponding to the above pressure is obtained by a simple integration with respect to D, and gives

$$W = (64kT\rho_\infty \gamma^2/\kappa)e^{-\kappa D} = 0.0482\,[\text{NaCl}]^{1/2}\gamma^2 e^{-\kappa D} \quad \text{J m}^{-2}, \quad (12.47)$$

where in both of the above equations [NaCl] is in M and the values are for aqueous solutions at 298 K.

Using the Derjaguin approximation, Eq. (10.16), we may immediately write the expression for the force F between two spheres of radius R as $F = \pi R W$, from which the interaction free energy is obtained by a further integration:

$$W = (64\pi kTR\rho_\infty \gamma^2/\kappa^2)e^{-\kappa D} = 4.61 \times 10^{-11}R\gamma^2 e^{-\kappa D} \quad \text{J}. \quad (12.48)$$

We see therefore that the electrostatic interaction between surfaces or particles decays exponentially with distance. The characteristic decay length is the Debye length.

At low surface potentials, below about 25 mV, all the above equations simplify to the following: For two planar surfaces,

$$P \approx 2\varepsilon\varepsilon_0 \kappa^2 \psi_0^2 e^{-\kappa D} = 2\sigma^2 e^{-\kappa D}/\varepsilon\varepsilon_0 \quad (12.49)$$

and

$$W \approx 2\varepsilon\varepsilon_0 \kappa \psi_0^2 e^{-\kappa D} = 2\sigma^2 e^{-\kappa D}/\kappa\varepsilon\varepsilon_0 \quad \text{(per unit area)}, \quad (12.50)$$

while for two spheres of radius R,

$$F \approx 2\pi R\varepsilon\varepsilon_0 \kappa \psi_0^2 e^{-\kappa D} = 2\pi R\sigma^2 e^{-\kappa D}/\kappa\varepsilon\varepsilon_0 \quad (12.51)$$

and

$$W \approx 2\pi R\varepsilon\varepsilon_0 \psi_0^2 e^{-\kappa D} = 2\pi R\sigma^2 e^{-\kappa D}/\kappa^2\varepsilon\varepsilon_0. \quad (12.52)$$

In the above ψ_0 and σ are related by $\sigma = \varepsilon\varepsilon_0\kappa\psi_0$, which, as we have seen, is valid for low potentials. These four equations are quite useful because they are valid for *all* electrolytes, whether $1:1$, $2:1$, $2:2$, $3:1$, or even mixtures, or so long as the appropriate Debye lengths are used as given by Eqs. (12.33)–(12.36). Thus, they are particularly suitable when divalent ions are present since the surface charge and potential *is* often low due to ion binding.

All the expressions so far derived for the interactions of two double layers are accurate only for surface separations beyond about one Debye length. At smaller separations one must resort to numerical solutions of the Poisson–Boltzmann equation to obtain the exact interaction potential (Verwey and Overbeek, 1948; Honig and Mul, 1971) for which there are no simple expressions. In addition there is the question of *charge regulation* at small separations (i.e., does the surface charge density remain constant as two surfaces come close together, or do some of the counterions bind to the surfaces thereby reducing σ). This affects the form of the interaction potential. At large distances, beyond $1/\kappa$, the question does not arise, and the interaction pressures and energies are well described by Eqs. (12.46)–(12.48), where ψ_0 and σ are the values appropriate for the isolated surfaces (at $D = \infty$). But at progressively smaller separations, as the counterion concentration at each surface increases, specific ion binding can occur as discussed earlier. If there is no binding, the surface charge density will remain constant, and in the limit of small D the number density of monovalent counterions between the two surfaces will approach a uniform value of $2\sigma/eD$. Thus, from Eq. (12.44) the limiting pressure in this case is

$$P = kT \sum_i \rho_{\mathrm{mi}} = 2\sigma kT/eD \tag{12.53}$$

or

$$W = (2\sigma kT/e)\log D + \text{constant}, \tag{12.54}$$

that is, as $D \to 0$ the pressure and energy become infinite. Note that this is the same limiting pressure as in the case of no bulk electrolyte (counterions only), as obtained in Section 12.9, and results purely from the limiting osmotic pressure of the "trapped" counterions.

If there is counterion binding as D decreases, the Poisson–Boltzmann equation must now be solved self-consistently by including the dissociation constants of the adsorbing ions (Section 12.12). The computations have been described by Ninham and Parsegian (1971), and a simple numerical algorithm has been given by Chan *et al.* (1980b).

The two main effects of a charge-regulating interaction can be summarized qualitatively as follows (see also Healy *et al.*, 1980):

(i)   if counterions adsorb as two surfaces approach each other the strength of the double-layer interaction is always less than that occurring at constant surface charge (Le Chatelier's principle), and

(ii)   in general, the interaction potential will lie between two limits, the one corresponding to the interaction at constant surface charge and the other at constant surface potential.

This is illustrated in Fig. 46, which shows the double-layer interaction potentials of two planar surfaces in 1:1 electrolyte. The curves are based on

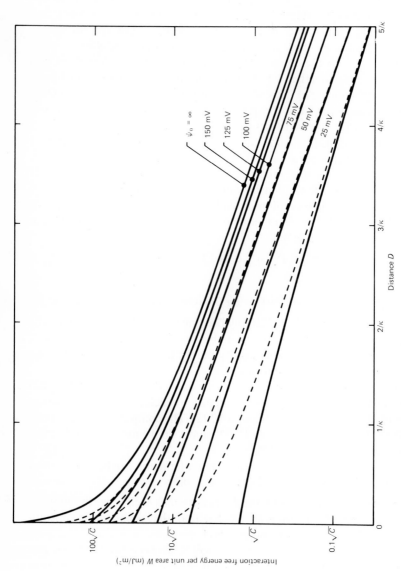

Fig. 46 Repulsive double-layer interaction for two planar surfaces in 1:1 electrolyte [exact solutions kindly computed by M. Sculley, R. Pashley, and L. White based on Ninham and Parsegian (1971)]. ψ_0 is the potential of the isolated surfaces and C the electrolyte concentration in M, which is related to the Debye length by $1/\kappa = 0.304/\sqrt{C}$ nm. Theoretically, the double-layer interaction must lie between the constant-charge and constant-potential limits. ——, constant charge; ——, constant potential.

exact numerical solutions and show the theoretical limits of the constant charge and constant potential interaction in each case. The figure may be used for reading off the interaction energy of any 1:1 electrolyte at any desired concentration. The constant potential curves of Fig. 46 compare reasonably well with the approximate expression of Eq. (12.47) even at small separations, and especially when ψ_0 is between 50 and 100 mV. However, the accurately computed constant charge interaction curves are always well above the values predicted by the approximate expression at small separations. Approximate expressions for interactions at constant surface charge were given by Gregory (1973).

Finally, it is worth mentioning that for two surfaces of different charge densities or potentials the interaction energy can have a maximum at some finite distance (usually well below $1/\kappa$) (i.e., the surfaces can *attract* each other at small separations). Approximate equations for the interactions of two surfaces of unequal but constant potentials were given by Hogg et al. (1966) and, in a different form, by Parsegian and Gingell (1972), and for unequal charges by Gregory (1975).

12.17. van der Waals and double-layer forces acting together: the DLVO theory

The total interaction between any two surfaces must also include the van der Waals attraction. Now, unlike the double-layer interaction, the van der Waals interaction potential is largely insensitive to variations in electrolyte concentration and pH, and so may be considered as fixed for any particular solute–solvent system (but see comments in Section 17.1). Further, the van der Waals attraction must always exceed the double-layer repulsion at small distances since it is a power law interaction (i.e., $W \propto -1/D^n$), whereas the double-layer interaction energy remains finite or rises much more slowly as $D \to 0$. Figure 47 shows schematically the various types of interaction potentials that can occur between two surfaces under the combined action of these two interactions. Depending on the electrolyte concentration and surface charge density or potential one of the following may occur:

(i) For highly charged surfaces in dilute electrolyte (i.e., long Debye length), there is a strong long-range repulsion that peaks at some distance (usually between 1 and 4 nm) at the *energy barrier*. This is illustrated in Fig. 47a.

(ii) In more concentrated electrolyte solutions there is a significant *secondary minimum* (usually beyond 4 nm) before the energy barrier (Figs. 47b and 78). The potential energy minimum at contact is known as the *primary minimum*.

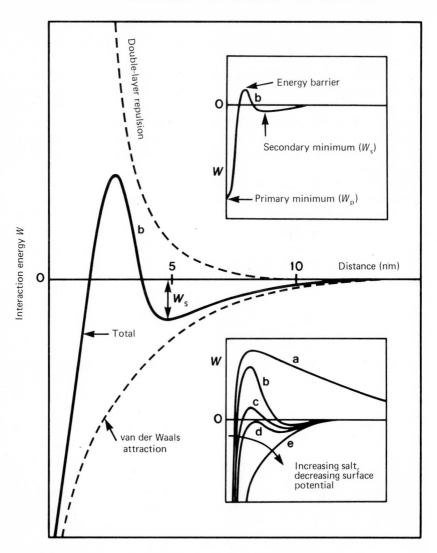

Fig. 47 Schematic energy versus distance profiles of DLVO interaction.
(a)   Surfaces repel strongly; small colloidal particles remain "stable."
(b)   Surfaces come into stable equilibrium at secondary minimum if it is deep enough; colloids remain "stable."
(c)   Surfaces come into secondary minimum; colloids coagulate slowly.
(d)   Surfaces may remain in secondary minimum or adhere; colloids coagulate rapidly.
(e)   Surfaces and colloids coalesce rapidly.

(iii)   For surfaces of low charge density, the energy barrier will always be much lower (Fig. 47c), and above some concentration of electrolyte (the *critical coagulation concentration*) it will fall below the $W = 0$ axis (Fig. 47d). When this happens small colloidal particles in solution rapidly come together or "coagulate" (cf. Fig. 28i). This important phenomenon forms the basis of the so-called DLVO theory of colloidal stability (Derjaguin and Landau, 1941; Verwey and Overbeek, 1948; Shaw, 1970; Hiemenz, 1977).

(iv)   As the surface charge approaches zero the interaction curve approaches the pure van der Waals curve, and the two surfaces now attract each other at all separations (Fig. 47e).

The main factor inducing two surfaces to come into contact is the lowering of their surface charge or potential, brought about by increased ion binding (e.g., of Ca^{2+} or H^+ for negatively charged surfaces). However, surfaces of high charge can still come close together, into a secondary minimum if not necessarily into contact, by raising the salt concentration.

It is clear that one must have a fairly good idea of the charging process occurring at a surface before attempting to understand its double-layer interactions.

12.18. Experimental measurements of double-layer forces

Figure 48 shows detailed experimental results of direct force measurements between two mica surfaces in dilute 1:1 and 2:1 electrolyte solutions where the Debye length is large, thereby allowing accurate comparison with theory to be made at distances as small as $0.02\kappa^{-1}$. The theoretical DLVO force laws are shown by the continuous curves. The agreement is remarkably good at all separations, even down to 2% of the Debye length, and indicates that the DLVO theory is basically sound. One may also conclude that the dielectric constant of water must be the same as the bulk value even at surface separations as small as 2 nm, since otherwise significant deviations from theory would have occurred. (Hamnerius *et al.*, 1978, showed that the dielectric constant of water remains unchanged even for 1-nm films.) The surface potentials ψ_0, inferred from the magnitude of the double-layer forces at large distances, agree within 10 mV with those measured independently on isolated mica surfaces (Lyons *et al.*, 1981). Further, the surface charge density corresponding to these potentials is typically $1e$ per 60 nm^2. Thus, at separations below about 8 nm the surfaces are actually closer to each other than the mean distance between the surface charges, and yet the double-layer forces still behave as if the surface charges were smeared out. The reason for this will become clear in the next section.

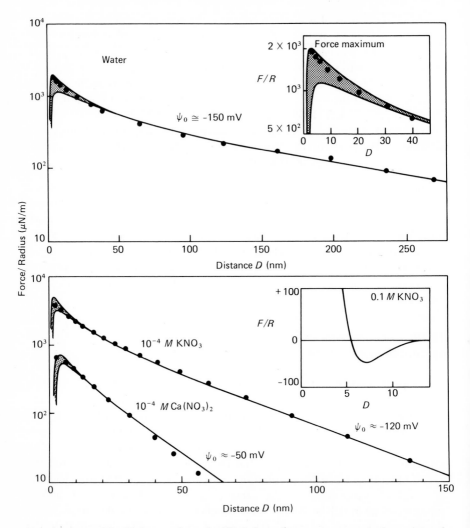

Fig. 48 Measured double-layer and van der Waals forces between two curved mica surfaces of radius R (1 cm) in water and in dilute $\sim 10^{-4}$ M KNO_3 and $\sim 10^{-4}$ M $Ca(NO_3)_2$ solutions. The continuous curves are the theoretical DLVO forces (using a Hamaker constant of $A = 2.2 \times 10^{-20}$ J) showing the constant charge and constant potential limits. Theoretically, we expect the interactions to fall between these two limits. (Note that for this geometry the Derjaguin approximation gives $F/R = 2\pi W$, where W is the corresponding interaction energy per unit area between two planar surfaces, as plotted in Fig. 46.) The inset in the lower part of the figure is the measured force in concentrated 0.1 M KNO_3 showing the emergence of a secondary minimum (cf. Fig. 78). [From Israelachvili and Adams (1978), Pashley (1981a), and Israelachvili (1982).]

Other measurements of double-layer forces have been carried out in various 1:1 and 2:1 electrolytes (Pashley, 1981a,b; Pashley and Israelachvili, 1984), between surfactant bilayers (Pashley and Israelachvili, 1981), across soap films as in Fig. 28e (Derjaguin and Titijevskaia, 1954; Lyklema and Mysels, 1965; Donners et al., 1977), between glass surfaces (Rabinovich et al., 1982; Peschel et al., 1982), as well as in nonaqueous polar liquids (Christenson and Horn, 1983). The results are invariably in good agreement with theory (though additional solvation forces are often observed at separations below 1–5 nm, depending on the system).

It is perhaps surprising that measured double-layer forces are so well described by a theory that, unlike van der Waals force theory, contains a number of fairly drastic assumptions (e.g., the assumed smearing out of discrete surface charges, that ions can be considered as point charges, and that the Poisson–Boltzmann equation remains valid even at fairly high concentrations). Any deviations in the forces from those expected from the DLVO theory can usually be traced to the existence of some other force (e.g., solvation force) and not because of a breakdown in the DLVO interaction. These short-range solvation forces are, of course, very important, especially in determining the coagulation of colloidal particles and in biological systems where short-range interactions are of paramount importance. Their consideration forms a large part of the rest of this book.

12.19. Effects of discrete surface charges and dipoles

The charge on a surface is obviously not uniformly smeared out over the surface, as has been implicit in all the equations derived so far. For a surface of typical potential of 75 mV in a 1 mM NaCl solution, the surface charge density, as given by the Grahame equation, is $\sigma = 0.0075$ C m^{-2}, which corresponds to only one charge per 21 nm^2 or 2100 Å^2! In 0.1 M NaCl the same potential implies 1e per 2 nm^2. Thus, the charges on real surfaces are typically 1–5 nm apart from each other on average. What effect does this have on the electrostatic interaction between two surfaces?

Let us consider a planar square lattice of like charges q as shown in Fig. 49a. If d is the distance between any two neighbouring charges, then the mean surface charge density is $\sigma = q/d^2$, and if this charge were smeared out, the electric field emanating from the surface would be $E_z = \sigma/2\varepsilon\varepsilon_0$. What, then, is the field of a discrete surface lattice having the same mean overall charge density? To compute this field one must sum the contributions from all the charges. The resulting slowly converging series can be turned into a rapidly converging series by using a mathematical technique known as the *Poisson summation formula* (Lighthill, 1970). If x and y are the coordinates in the plane

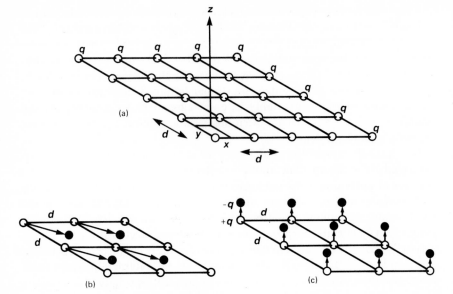

Fig. 49 Sections of infinite lattices of charges and dipoles.

relative to any charge as the origin (Fig. 49a), the field E_z along the z direction is given by the series (Lennard-Jones and Dent, 1928)

$$E_z = \frac{\sigma}{2\varepsilon\varepsilon_0}\left[1 + 2\left(\cos\frac{2\pi x}{d} + \cos\frac{2\pi y}{d}\right)e^{-2\pi z/d} + \cdots\right], \qquad (12.55)$$

where the higher-order terms decay much more rapidly with distance z. The first term is the same as that of a smeared-out surface charge. The second term is interesting, for it shows that the excess field decays away extremely rapidly, with a decay length of $d/2\pi$ (e.g., about 0.3 nm for charges 2 nm apart). Thus, at $z = \frac{1}{2}d$ the electric field is at most 17% different from that of the smeared-out field, while at $z = d$ it has reached 99.3% of the smeared-out value! It is for this reason that the smeared-out approximation works so well in considering the electrostatic interactions between charged surfaces.

If charges of opposite sign are now added at the center of each square, as in Fig. 49b, the net surface charge density becomes zero. By superimposing the fields of the positive and negative charge lattices using Eq. (12.55) it is easy to show that the electric field opposite a positive charge (at $x = 0$, $y = 0$) is

$$E_z = +(4q/\varepsilon\varepsilon_0 d^2)e^{-2\pi z/d} + \cdots, \qquad (12.56)$$

while opposite a negative charge (at $x = \frac{1}{2}d$, $y = \frac{1}{2}d$) it is

$$E_z = -(4q/\varepsilon\varepsilon_0 d^2)e^{-2\pi z/d} + \cdots. \qquad (12.57)$$

Note that this geometry is equivalent to a dipolar or zwitterionic lattice whose dipoles, of length $d/\sqrt{2}$ and surface density $1/d^2$, are lying parallel to the surface. For dipoles arrayed perpendicular to the surface (Fig. 49c), the above two equations become replaced by $E_z \approx \pm(2q/\varepsilon\varepsilon_0 d^2)e^{-2\pi z/d}$. This procedure can be readily extended to other lattices including three-dimensional ionic crystals. The end result is always that the field decays very rapidly to zero.

If a second lattice of vertical dipoles is brought up to the first, the Coulombic interaction pressure between the two surfaces at a separation D will vary between $\pm(2q^2/\varepsilon\varepsilon_0 d^4)e^{-2\pi z/D}$ depending on whether the approaching dipoles are exactly opposite each other (repulsion) or in between (attraction). The pressure is anyway very small, and in reality since surface dipoles will not be on a perfect lattice but distributed randomly or moving about (e.g., zwitterions on a lipid bilayer surface) the net pressure will average to zero. A similar result is obtained if the dipoles are lying in the plane of each surface, as in Fig. 49b. However, Jönsson and Wennerström (1983) pointed out that the interaction between each dipole and its image reflected by the other surface is always present and that for two dipolar surfaces interacting across water the resulting pressure can be large and repulsive, decaying either exponentially or with the inverse fourth power of the separation depending on the lateral correlation between dipoles. They proposed that this purely coulombic interaction is responsible for the short-range repulsive "hydration forces" observed between zwitterionic lipid bilayers in water at small distances (cf. Fig. 41).

Chapter 13

Solvation and Steric Interactions

13.1. Molecular ordering at interfaces

The theories of van der Waals and double-layer forces, discussed in the previous two chapters, are both *continuum theories* wherein the intervening solvent is characterized solely by its bulk properties such as refractive index n, dielectric constant ε, and density ρ. We have already seen in Chapters 7 and 8 that at small separations, below a few molecular diameters, the discrete molecular nature of a solvent cannot be ignored and that the short-distance intermolecular pair potential can be quite different from that expected from continuum theories. In particular we saw that in general the liquid density profiles and interaction potentials in liquids oscillate with distance, with a periodicity close to the molecular size and with a range of a few molecular diameters. These short-distance interactions are usually referred to as *solvation* or *structural forces*, and to understand how they act between surfaces we must first consider the way solvent molecules order themselves at an isolated surface, then how this ordering becomes modified in the presence of a second surface and how this determines the short-range solvation interaction between two surfaces in liquids. The solvation phenomenon at surfaces is in principle no different from that occurring around a solute molecule, or around another solvent molecule, which—as previously described—is determined primarily by the geometry of molecules and how they can pack around a constraining boundary [for two mainly theoretical reviews see Nicholson and Parsonage (1982) and Rickayzen and Richmond (1985)].

Theoretical work and Monte Carlo computer simulations indicate that while liquid density oscillations are not expected to occur at a vapour–liquid or liquid–liquid interface (Fig. 50a), a very different situation arises at a solid–

194

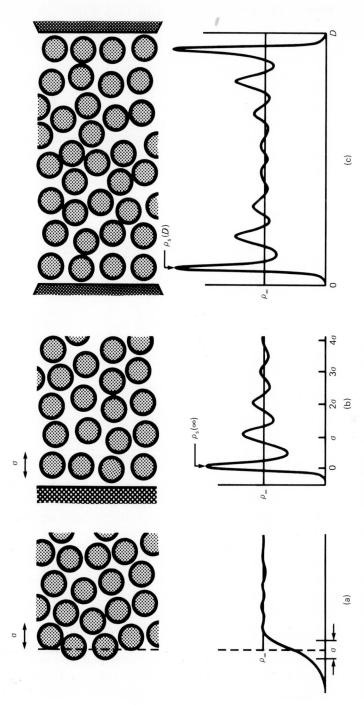

Fig. 50 (a)   Liquid density profile at a vapour–liquid interface. ρ_∞ is the bulk liquid density.

(b)   Liquid density profile at an isolated solid–liquid interface. $\rho_s(\infty)$ is the "contact" density at the surface, which increases with the strength of the attractive potential between the wall and liquid molecules.

(c)   Liquid density profile between two walls a distance D apart. The contact density $\rho_s(D)$ is a function of D as illustrated in Fig. 51.

liquid interface (Fig. 50b). Here, attractive interactions between the wall and liquid molecules, and the geometric constraining effect of the wall on these molecules, give rise to density oscillations, or "structure," extending several molecular diameters into the liquid (Abraham, 1978; Rao et al., 1979). The constraining effect of two solid surfaces is much more dramatic (Fig. 50c). Even in the absence of any wall–liquid interaction the liquid molecules must reorder themselves so as to be accommodated between the two walls, and the variation of this ordering with separation D gives rise to the solvation force between the two surfaces.

13.2. Origin of solvation forces

In Section 12.7 we saw that the repulsive electrostatic double-layer pressure between two surfaces separated by a solvent containing ions (which do not interact with, i.e., bind to, the surfaces) is given by the contact value theorem, Eq. (12.18):

$$P(D) = kT[\rho_s(D) - \rho_s(\infty)], \qquad (13.1)$$

where ρ_s is the ionic density at each surface. Equation (13.1) also applies to solvation forces so long as there is no interaction between the walls and liquid molecules, where ρ_s is now the density of liquid molecules at each surface (Fig. 50b,c). Thus, a solvation force arises once there is a change in the liquid density at the surfaces as they approach each other. For noninteracting "hard walls," this is brought about by changes in the molecular packing as D varies, as illustrated in Fig. 51. Here we see that ρ_s will be high only at surface separations that are multiples of σ but must fall at intermediate separations. At large separations, as ρ_s approaches the value for isolated surfaces $\rho_s(\infty)$, the solvation pressure approaches zero. The resulting variation of the solvation pressure with distance is also shown schematically in Fig. 51; like the density profile, it is an oscillatory function of distance of periodicity roughly equal to σ and with a range of a few molecular diameters.

So far we have looked at the simplest idealized case of "hard-sphere molecules" between two "hard walls" without considering the effects of specific interactions between the wall and liquid molecules or between the liquid molecules themselves. Let us consider these in turn.

First, any attractive interaction between the surfaces and the liquid molecules leads to a denser packing of molecules at the walls (Abraham, 1978; Snook and van Megen, 1979) and thus to higher ρ_s values and a stronger solvation interaction. A number of theoretical studies and Monte Carlo simulations of Lennard-Jones liquids (Chan et al., 1980a; van Megen and Snook, 1979, 1981; Snook and van Megen, 1980, 1981; Grimson et al., 1980)

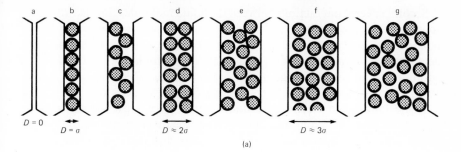

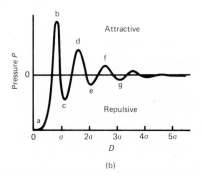

Fig. 51 (a)   Same as in Fig. 50c, showing how the molecular ordering changes as the separation D changes. Note that the density of liquid molecules in contact with the surfaces $\rho_s(D)$ varies between maxima and minima.
(b)   Corresponding solvation pressure (schematic) as given by Eq. (13.1). In general the oscillatory solvation force need not be symmetrical about the $P = 0$ axis.

and water (Christou *et al.*, 1981) between surfaces with which the liquid molecules interact via some form of the Mie potential have invariably lead to an oscillatory solvation force at separations below a few molecular diameters (Rickayzen and Richmond, 1985). It is important to note that once the solvation zone of two surfaces overlap, the mean density ρ of the intervening liquid is no longer the same as that of the bulk liquid. And since the van der Waals–Lifshitz interaction depends on both n and ε, which in turn depend on ρ, we must conclude that *van der Waals and oscillatory solvation forces are not additive.* Indeed, it is more correct to think of the solvation force as *the* van der Waals force at small separations with the molecular properties of the medium taken into account. It is also important to appreciate that solvation forces do not arise simply because liquid molecules tend to lie in semiordered layers at surfaces, but because of the disruption of this ordering during the approach of a second surface. The two effects are of course related: the greater the tendency towards stratification, the greater the solvation force, but the distinction between them should always be borne in mind.

Second, specific interactions between the liquid molecules can also affect the solvation force. In particular, it is believed that for water, long-range dipole polarization and cooperative H-bonding effects can lead to a monotonically decaying repulsive "hydration" or attractive "hydrophobic" interaction, in addition to the shorter-range oscillatory interaction (Marčelja and Radić, 1976; Marčelja et al., 1977; Gruen and Marčelja, 1983; Jönsson, 1981). Note that the long-range electrostatic double-layer force must strictly also be thought of as a solvation force, arising from the different concentrations of solute ions at surfaces from those in the bulk and their variation as two surfaces approach each other.

13.3. Measurements of solvation forces in nonaqueous liquids

While theoretical work on real systems is still in its infancy there is a rapidly growing literature on experimental measurements and other phenomena associated with solvation forces. The simplest systems so far investigated have involved direct measurements of these forces between molecularly smooth surfaces in organic liquids. Figure 52 shows the results obtained by Horn and Israelachvili (1981) for two mica surfaces in octamethylcyclotetrasiloxane (OMCTS) $[(CH_3)_2 SiO]_4$—an inert liquid whose nonpolar molecules are quasi-spherical with a mean molecular diameter of about 0.9 nm. Subsequent measurements by Christenson (1983) and Christenson and Horn (1983) on a variety of different liquids have lead to qualitatively similar solvation force laws whose features may be summarized as follows:

(i)   In liquids such as benzene C_6H_6, cyclohexane C_6H_{12}, CCl_4, and OMCTS whose molecules are fairly rigid, the solvation force oscillates with distance, varying between attraction and repulsion, with a periodicity equal to the mean molecular diameter σ (within a few percentage points of the diameters obtained from x-ray, gas solubility, and diffusion data).

(ii)   The peak-to-peak amplitudes of the oscillations show a roughly exponential decay with distance with a characteristic decay length of 1.2 to 1.7σ.

(iii)   The solvation force can exceed the van der Waals force at separations below 6 to 10 molecular diameters, and merges with the continuum van der Waals force at larger separations (cf. Fig. 52).

(iv)   The depth of the potential well at contact ($D = 0$) corresponds to an interaction energy that is surprisingly close to the value expected from the continuum Lifshitz theory of van der Waals forces. For example, the Lifshitz Hamaker constant for the mica–OMCTS–mica system is about $1.35 \times$

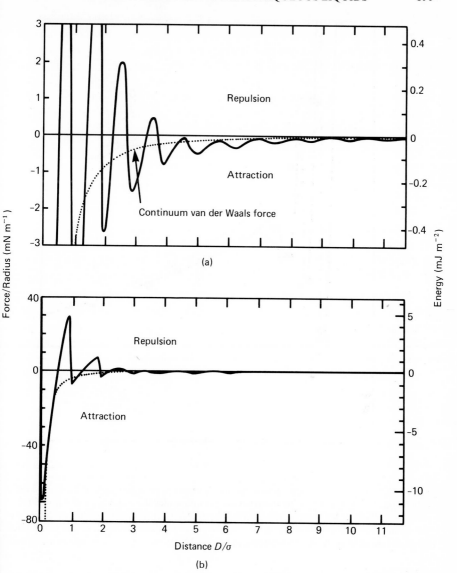

Fig. 52 (a)   Measured force F between two curved mica surfaces of radius R (≈ 1 cm) in OMCTS ($\sigma \approx 0.9$ nm) at 22°C. For comparison, the theoretical continuum van der Waals force is also shown.

(b)   The full force law plotted on a reduced scale. The right hand ordinate gives the corresponding interaction energy per unit area of two flat surfaces according to the Derjaguin approximation: $W = F/2\pi R$. (From Horn and Israelachvili, 1981.)

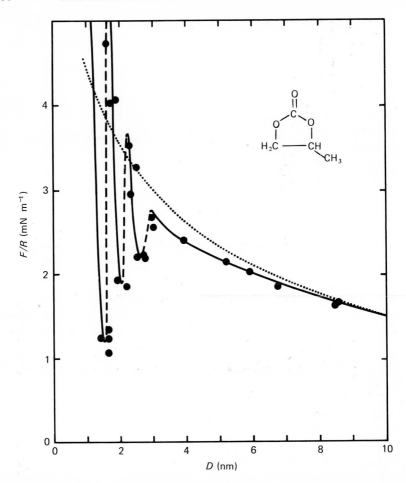

Fig. 53 Measured force between two curved mica surfaces in propylene carbonate ($\sigma \approx 0.5$ nm) containing 10^{-4} M electrolyte (tetraethylammonium bromide). Beyond 8 nm and out to 50 nm (not shown) the force law is accurately described by double-layer theory (dotted line). Below 8 nm the van der Waals force reduces the net force below the purely double-layer curve, and below 3nm the force is oscillatory. (From Christenson and Horn, 1983).

10^{-20} J. Using Eq. (11.26) with a cut-off contact distance of $D_0 = 0.17$ nm (Section 11.8) we obtain $W \approx A/12\pi D_0^2 \approx 12$ mJ m^{-2} for the adhesion energy at contact, which may be compared with the value of $W \approx 11$ mJ m^{-2} obtained from the measured adhesion force (Fig. 52b).

(v)   Solvation forces are not strongly temperature dependent and so cannot be viewed as a surface-induced "prefreezing" of liquids.

(vi)   For flexible molecules such as *n*-octane and 2,2,4-trimethylpentane, whose internal bonds can rotate freely, there is no long-range structure, and the solvation force does not extend beyond two to four molecules.

(vii)   In polar liquids such as acetone (dipole moment 2.85 D) the solvation force is not qualitatively different from that of nonpolar liquids. In polar and hydrogen-bonding liquids of high dielectric constant such as propylene carbonate ($\varepsilon = 65$), methanol ($\varepsilon = 33$), and ethylene glycol ($\varepsilon = 41$) containing dissolved ions, there is a double-layer force at large distances, but at smaller distances the oscillatory solvation force, rather than the continuum van der Waals force, dominates the interaction (Fig. 53).

(viii)   The presence of even trace amounts of water can have a dramatic effect on the solvation forces in nonpolar liquids, especially if there is a preferential adsorption of water onto surfaces that leads to a disruption of the molecular ordering in the first few layers (see also Chapter 14).

Horn and Israelachvili (1981) discussed the implications of solvation forces between hard surfaces in nonpolar organic liquids (e.g., oils) for adhesion, lubrication, and interparticle interactions in colloidal dispersions. For liquid–liquid and liquid–vapour interfaces, we may expect solvation forces to be much weaker since, as was illustrated in Fig. 50, there is little or no structuring expected at such interfaces. It is not surprising therefore that measurements of the van der Waals forces across thin liquid films of helium and alkanes on surfaces are well described by the Lifshitz theory down to film thickness of 1 nm (Section 11.5).

13.4. Solvation forces in aqueous systems: hydration forces

There are many aqueous systems where DLVO theory fails and where additional short-range (< 5 nm) exponentially repulsive solvation forces are observed, commonly referred to as *hydration, structural,* or *hydrophilic forces.* The origin and nature of these forces has long been controversial especially in the colloid literature (Derjaguin and Churaev, 1977; Eagland, 1975), but in recent years there has been much progress in their understanding and characterisation. Hydration forces arise whenever water molecules bind to surfaces containing hydrophilic groups (i.e., certain ionic, zwitterionic, or H-bonding groups), and their strength depends on the energy needed to disrupt the ordered water structure and ultimately dehydrate two surfaces as they approach each other (van Olphen, 1977).

Certain clays, surfactant soap films, and lipid bilayers swell spontaneously in aqueous solutions, and silica dispersions and other colloidal particles sometimes remain "stable" in very high salt. And yet in these systems one would expect the surfaces or particles to remain in strong adhesive contact or coagulate in a primary minimum if DLVO forces alone were operative. Very strong short-range repulsive forces have been measured across soap films composed of the cationic surfactant hexadecyltrimethylammonium bromide (HTAB) and across bilayers composed of the zwitterionic lipid phosphatidylcholine (lecithin), shown in Fig. 54a,b. In these systems the hydration forces are due to the strongly hydrophilic $-N(CH_3)_3^+$ and phosphate— O PO_2^-O—groups (see Table XIV) that are present on the surfaces. In the case of glass and silica (Fig. 54c,d) the hydration forces arise from the strongly H-bonding surface hydroxyl ($-OH$) groups. These forces are also believed to be responsible for the unexpectedly thick wetting films of water on silica, discussed in Section 12.8.

In an extensive series of experiments Pashley (1981a,b) and Pashley and Israelachvili (1984) found that the interaction between molecularly smooth mica surfaces in dilute electrolyte solutions obeys the DLVO theory (cf. Fig. 48). However, at higher salt concentrations, specific to each electrolyte, hydrated cations bind to the negatively charged surfaces and give rise to a repulsive hydration force (shown in Fig. 55 for K^+) due to the energy needed to dehydrate the bound cations, which presumably retain some of their water of hydration on binding. This conclusion was arrived at after noting that the strength and range of the hydration forces increased with the hydration number of the cations (cf. Table V) in the order $Mg^{2+} > Ca^{2+} > Li^+ \sim Na^+ > K^+ > Cs^+$. In acid solutions, when only protons bind to the surfaces, no hydration forces were observed, and the measured force laws were very close to those expected from DLVO theory.

These results show that hydration forces can be modified or *regulated* by exchanging ions of different hydrations on surfaces. Such regulated hydration effects also occur at other surfaces. For example, the force between two mercury surfaces obeys DLVO theory in various electrolyte solutions, but the surfaces fail to coalesce once ionic species (e.g., I^-) specifically bind to the surfaces at higher concentrations (Usui *et al.*, 1967; Usui and Yamasaki, 1969). Regarding hydration regulation in colloidal dispersions, the effects of different electrolytes on the hydration forces between colloidal particles can determine whether they will coagulate or not. Figure 56 shows the experimentally determined regions of stability and instability of amphoteric polystyrene latex particles in $CsNO_3$ and KNO_3 solutions. In concentrated $CsNO_3$ solutions Cs^+ binds to the surfaces at high pH (where there is no competition from protons). But the hydration forces are expected to be weak, and the experimental regions are indeed explicable by DLVO theory. However, in concentrated KNO_3 and $LiNO_3$ solutions the particles remain stable, even at the isoelectric point, because of the hydration forces arising from the binding of the more hydrated K^+ and Li^+ ions.

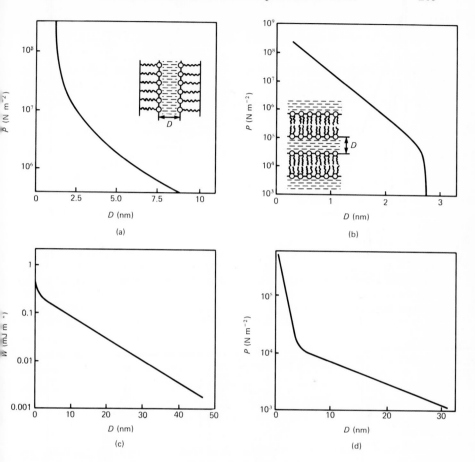

Fig. 54 Experimental measurements of repulsive hydration forces.

(a) Across a soap film of the cationic surfactant HTAB: $C_{16}H_{33}N(CH_3)_3{}^+Br^-$ (From Clunie *et al.*, 1967. Reprinted by permission from *Nature*, Vol. 216, pp. 1203–1204. © 1967 MacMillan Journals Limited.). A similar strong short-range repulsion below ~1.2 nm was measured between two HTAB bilayers by Pashley and Israelachvili (1981).

(b) Between bilayers of the uncharged lipid egg-phosphatidylcholine (lecithin). Below 2.5 nm the repulsive pressure is exponential with a characteristic decay length of about 0.26 nm. (From Parsegian *et al.*, 1979.) Similar results were obtained for a variety of other neutral lipids by Lis *et al.* (1982).

(c) Between two glass fibres. (From Rabinovich *et al.*, 1982.)

(d) Between two silica surfaces. (From Peschel *et al.*, 1982.) In both (c) and (d) the force is pure double layer down to 4–5 nm below which there is an additional exponentially repulsive hydration force (of decay length ~1 nm) instead of the attractive van der Waals force expected from DLVO theory.

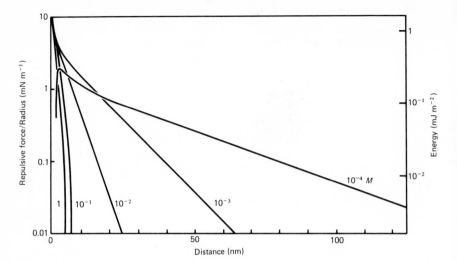

Fig. 55 Measured forces between curved mica surfaces of radii ~ 1 cm in KNO_3 or KCl solutions. In 10^{-4} M the results follow the theoretical DLVO force law. At 10^{-3} M and higher concentrations there is an additional short-range repulsion below about 4 nm of exponential decay length ~ 1 nm, the same as that observed between glass and silica surfaces. The right-hand ordinate gives the interaction energy between two flat surfaces according to the Derjaguin approximation. (From Israelachvili and Pashley, 1982a; Pashley, 1981a,b.)

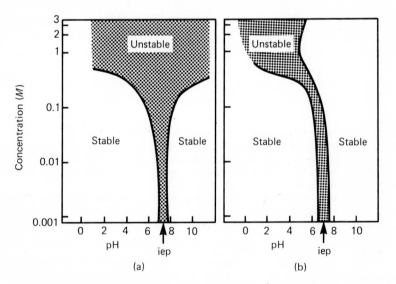

Fig. 56 Domains of stability and instability (coagulation) of a dispersion of amphoteric particles whose surfaces contain —COO^- and —NH_3^+ groups.
(a) In $CsNO_3$ solutions the behaviour is "ideal"—that expected from DLVO theory.
(b) In KNO_3, and to an even greater extent in $LiNO_3$, it is not—the expected coagulation does not occur in high salt above the isoelectric point (iep = 7.2). (From Healy *et al.*, 1978.)

From the discussion so far we can infer that hydration forces are not of a simple nature. First, a distinction must be made between hydration forces that occur between surfaces that are intrinsically hydrophilic (e.g., silica and lecithin) and those whose hydrophilic groups can be regulated by, for example, ion exchange. Intrinsically hydrophilic surfaces cannot be coagulated by simply adding more salt or changing the pH. Second, for surfaces whose hydration forces can be regulated, their effect, when coupled with the DLVO interaction, can be quite complex. In particular we may note that the commonly observed coagulation of negatively charged particles at low pH may be due not only to (i) the reduction or neutralization of the surface charge and a shorter Debye screening length (DLVO theory), but also to (ii) the elimination of hydration forces as protons replace bound hydrated ions on surfaces.

How do exponentially repulsive hydration forces arise? Why are they not oscillatory? Theoretical work in this area is only just beginning. Monte Carlo simulations by Jönsson (1981) and Christou et al. (1981) suggest oscillatory behaviour, while other theoretical studies (Marčelja and Radić, 1976; Marčelja et al., 1977; Jönsson and Wennerström, 1983) suggest a monotonic exponential repulsion, possibly superimposed on an oscillatory profile. Israelachvili and Pashley (1983) found that while the hydration force between two mica surfaces is overall repulsive below about 4 nm (Fig. 55), it is not always monotonic below about 1.5 nm but exhibits oscillations of mean periodicity 0.25 ± 0.03 nm, roughly the diameter of the water molecule (Fig. 57). In particular, they observed that the first three minima at $D = 0$, 0.28, and 0.56 nm occur at negative energies, a result that rationalizes observations on clay systems: Clay platelets such as montmorillonite repel each other increasingly strongly down to separations of ~ 2 nm (Viani et al., 1984). However, the platelets can also stack into stable aggregates with water interlayers of typical thickness 0.25 and 0.55 nm between them (Del Pennino et al., 1981). In chemistry we would refer to such structures as stable hydrates of fixed stoichiometry, while in physics we may think of them as experiencing oscillatory force.

It is probable that the short-range hydration force between smooth rigid surfaces (e.g., mineral surfaces such as mica) is always oscillatory, due to the ordered layering of water molecules bound to hydrated surface groups. This may be superimposed on a monotonically repulsive profile due to image interactions (Jönsson and Wennerström, 1983) and/or to a decaying co-operative H-bonding (polarization) interaction away from the surfaces (Marčelja and Radić, 1976; Gruen and Marčelja, 1983). It is also likely that for rough (e.g., silica) or fluid (e.g., soap film and bilayer) surfaces, the oscillations are smeared out resulting in a purely monotonic repulsion, as has so far always been observed (Fig. 54). It is clear, however, that the situation in water is

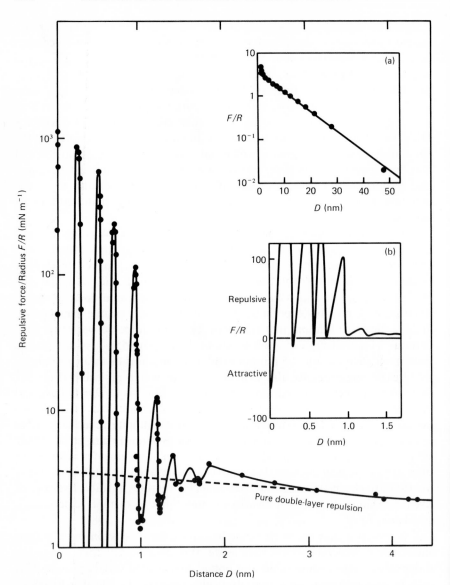

Fig. 57 Detailed measurements of forces between two curved mica surfaces in 10^{-3} M KCl at very small separations.

Inset (a): Force law at large separations, beyond 2 nm, where the solid line is the theoretical double-layer force for $\psi_0 = -78$ mV (cf. Fig. 55).

Inset (b): Force law at small separations, below 2 nm, plotted on a linear scale. (From Israelachvili and Pashley, 1983. Reprinted by permission from *Nature*, Vol. 306, No. 5940, p. 249. © 1983 Macmillan Journals Limited.)

governed by much more than the simple molecular packing effects that seem to dominate the interactions in nonaqueous liquids. In Part III we shall continue to consider the role of hydration and hydrophobic forces in the formation and interactions of micelles, bilayers, biological membranes, and other macro-molecular structures.

13.5. Hydrophobic forces

At any hydrophobic surface there is no binding of water molecules, and the oscillatory structure will be weak, especially at fluid hydrocarbon–water interfaces. But the orienting effects of cooperative H-bonding interactions between water molecules remain, giving rise to solvation effects that are believed to be quite different from those occurring at hydrophilic surfaces, even though at present the exact orientations of water molecules at hydrophilic or hydrophobic surfaces are unknown.

The strongly attractive solvation interaction between hydrophobic surfaces and groups in water is known as the hydrophobic interaction, which we examined in some detail in Chapter 8. There we saw that the orientation of water molecules adjacent to hydrophobic surfaces is entropically un-favourable and that two such surfaces therefore attract each other, since by coming together the entropically unfavoured water is ejected into bulk and the total free energy is thereby reduced.

So far there has been little experimental work done on the hydrophobic interaction potential between surfaces. Israelachvili and Pashley (1982b) measured these forces in the range 0–10 nm between two mica surfaces covered by HTAB monolayers (CH_3 and CH_2 groups exposed) in water and in various electrolyte solutions. They found that at surface separations below about 6 nm the hydrophobic attraction becomes much stronger than the expected van der Waals attraction and that it decays exponentially with distance with a decay length of about 1 nm. The hydrophobic interaction between surfaces therefore acts over long range, which accounts for the rapid coagulation of hydrophobic particles in water and for the ease with which water films rupture on hydrophobic surfaces. In this latter case the van der Waals force across a water film is generally repulsive (Section 11.5), but this is more than offset by the attractive hydrophobic interaction between the surface–water and water–air interfaces.

One intriguing question concerns the nature of the solvation force in water when both hydrophilic *and* hydrophobic groups are present in close proximity on a surface (e.g., on a lipid bilayer surface). The long-range attractive forces measured between surfactant and lipid bilayers appear to be well described by the van der Waals-dispersion force (Pashley and Israelachvili, 1981; Parsegian *et al.*, 1979; Ohshima *et al.*, 1982), suggesting that the hydrophobic interaction is not additive but becomes neutralized when the local structure of water molecules is dominated by their interaction with nearby hydrophilic groups. However, the matter is complex, and we shall return to consider this and other aspects of both these interactions in Part III.

13.6. Interactions of polymer-covered surfaces: steric forces

A polymer is a macromolecule composed of many monomer units or *segments*. If all the monomer units are the same, it is called a *homopolymer*, if different, a *copolymer*. Examples of synthetic homopolymers are polyethylene oxide $[-CH_2-CH_2-O-]_n$, polyvinyl alcohol $[-CH_2-CH(OH)-]_n$, polyacrylamide $[-CH_2-CH(CONH_2)-]_n$, and polystyrene $[-CH_2-CH(C_6H_5)-]_n$. Proteins are copolymers of amino acids. The molecular weight M_0 of a monomer unit is typically between 50 and 100, while the total molecular weight M often exceeds 10^6. When in solution, a polymer assumes the shape of a random coil in the absence of segment–segment interactions in the solvent. The mean configuration of such an "unperturbed" random coil is shown schematically in Fig. 58a. Its root-mean-square radius is known as the *radius of gyration* R_g (Flory, 1953) and is related to the number of segments n and effective segment length l by

$$R_g = \frac{\sqrt{n}\,l}{\sqrt{6}} = \frac{l\sqrt{M/M_0}}{\sqrt{6}}. \tag{13.2}$$

As an example, if $l = 1.0$ nm and the segment molecular weight is $M_0 = 200$, then for a polymer of $M = 10^6$ we obtain $R_g \approx 29$ nm. However, the real volume of the molecule is only a small fraction of the volume occupied by the radius of gyration. For instance if the segment diameter is about the same as the segment length l, then the molecular volume is $\pi(l/2)^2 nl \approx 0.8\,nl^3$, while the volume encompassed by R_g is $4\pi R_g^3/3 \approx 0.3 n^{3/2} l^3$. The ratio of these volumes is $\sim 3\sqrt{n}$. Thus, for a polymer of 1000 segments, only about 1% of the random coil volume is actually occupied!

More generally, the value of R_g is given by

$$R_g = \frac{\alpha\sqrt{n}\,l}{\sqrt{6}}. \tag{13.3}$$

The *intramolecular expansion factor* α is unity in an *ideal solvent* (one in which there are no interactions between the segments). In a *good solvent* there is a

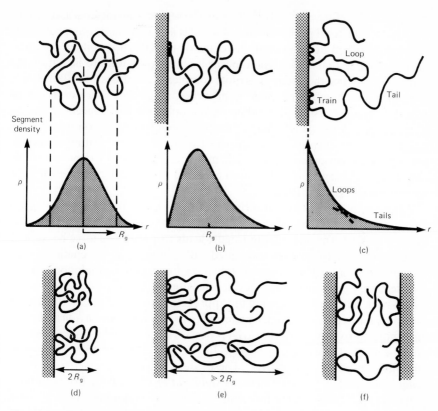

Fig. 58 (a)   Polymer in solution; (b) terminally grafted copolymer; (c) homopolymer; (d), (e) different states of adsorption; (f) bridging.

repulsion between the segments, the coil swells, and α exceeds unity. Polymers are completely soluble (miscible) in good solvents. In a *poor solvent* the segments attract each other, the coil shrinks, and α is less than unity. If the segment–segment attractions are particularly strong, due to ionic, van der Waals, or hydrophobic interactions, the coil loses all semblance of randomness and collapses or "folds" into a well-defined compact structure (e.g., proteins and DNA). A good solvent can often be made into a poor one by adding certain solutes or by lowering (or raising) the temperature below (or above) some critical value known as the theta temperature T_θ, at which point $\alpha = 1$.

Polymers absorb avidly on most surfaces, often reaching saturation adsorption at very low concentrations of a few parts per million in solution. If the radii of the adsorbed coils remain close to R_g (Fig. 58d), the area cover per molecule is therefore $\pi R_g^2 \approx \frac{1}{2}nl^2$. If again the segment diameter is assumed to be close to l, the total projected area of a coil is nl^2. Thus, as a rough guide, the amount adsorbed at full surface coverage of unperturbed coils is quantitatively similar to that which would occur if all the coils were to lie flat on the surface. In practice the situation is quite complex (Scheutjens and Fleer, 1982), and often much more polymer is adsorbed (Fig. 58e). Various types of adsorptions are possible, depending on the bulk polymer concentration, on whether the polymer is a homopolymer or a copolymer, and on whether the adsorption is via physical forces (physisorption) or by the grafting of specific anchor groups via chemical forces (chemisorption). These are depicted in Fig. 58b–e, as is the segment distribution expected theoretically for each type of adsorption (Hesselink, 1971).

When two polymer-covered surfaces approach each other they experience a force once the outer segments begin to overlap (i.e., once the separation is below a few R_g). This interaction usually leads to a repulsive force because of the unfavourable free energy associated with compressing the chains against the surfaces—a sort of complex solvation force that is analogous to the double-layer repulsion brought about when ions in the diffuse double layer are forced back onto surfaces during the approach of two surfaces. In the case of polymers this repulsion is usually referred to as the *steric interaction* and plays an important role in many industrial processes since colloidal dispersions can often be made stable to changes in electrolyte concentration and pH by small amounts of polymer additives. These are known as *protectives against coagulation*, and they lead to the *steric stabilization* of a colloid. Both synthetic polymers and biopolymers (proteins, gelatin) are commonly used in both nonpolar and polar solvents (e.g., in paints, ink, soils, emulsions, pharmaceutical dispersions) [for a review see Gregory (1978)].

Theories of steric interactions are complex and controversial (Hesselink et al., 1971; Vrij, 1976; Napper, 1977; Scheutjens and Fleer, 1982). Repulsive forces arise from the purely osmotic and finite-volume restricting effects of compressed coils plus the double-layer force in aqueous solutions. But at low coverage the van der Waals forces between two surfaces plus polymer bridging (Fig. 58f) lead to an attractive contribution (Gregory, 1978), and in a poor solvent the segments themselves always initially attract each other on overlapping.

Experimental measurements of steric interactions have so far been restricted to physisorbed homopolymers. Klein (1980, 1982) measured an initial attraction starting at $D \approx 3R_g$ followed by a steep repulsion as two polystyrene-covered mica surfaces approach each other in cyclohexane at $\sim 23°C$ (poor solvent conditions). The potential minima occurred at separations close to R_g (Fig. 59a). Israelachvili et al. (1984) found that the long-range attraction and shorter-range repulsion remains even above the theta

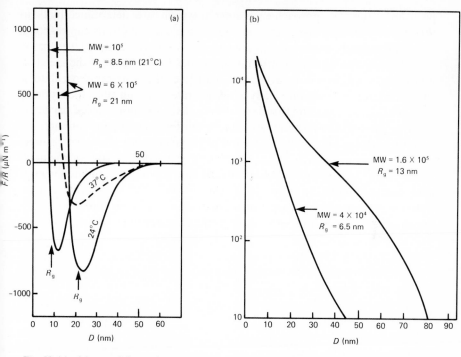

Fig. 59 (a)   Measured forces between two curved mica surfaces of radius R covered with adsorbed layers of polystyrene in cyclohexane below and above the theta temperature (34.5°C). (Adapted from Klein, 1980, 1982; Israelachvili *et al.*, 1984.)

(b)   Measured forces between mica surfaces covered with polyethylene oxide in aqueous 0.1 *M* KNO_3 solutions at 20°C where PEO is in a good solvent. (From Klein and Luckham, 1982. Reprinted by permission from *Nature*, Vol. 300, No. 5891, pp. 429–431. © 1982 Macmillan Journals Limited.)

temperature of $\sim 35°C$ where cyclohexane becomes a good solvent, though the potential minima now occur farther in (Fig. 59a).

A number of direct force measurements have also been reported in aqueous solutions (Cain *et al.*, 1978; Lyklema and van Vliet, 1978; Sonntag *et al.*, 1979; Israelachvili *et al.*, 1979; Klein and Luckham, 1982, 1984). In all these studies steric repulsions were measured at surface separations many times R_g (Fig. 59b) beyond which, depending on the conditions, the force can be attractive (due to bridging), as occurs in aqueous polyethylene oxide solutions (Israelachvili *et al.*, 1980c). Another common feature of these force measurements in aqueous solutions is the extreme sluggishness at which surfaces approach each other once their polymer layers begin to overlap. Now most theories of steric interactions are equilibrium theories, as are the theories of van der Waals and double-layer forces discussed earlier. When two colloid particles or surfaces approach each other at typical rates determined by their thermal motion both the electronic and the ionic distributions can respond sufficiently rapidly to ensure that the van der Waals and double-layer forces they experience *will* be the equilibrium forces (i.e., independent of the speed of approach, so long as it is not atypically high). Even the short-range oscillatory solvation forces in liquids are likely to relax at rates determined by the motions of the liquid molecules and therefore faster than the collision rates of particles. But one cannot always apply an equilibrium force law to the interactions between surfaces covered by long-chained polymers, since the time required for the solvent to be extruded between the entangled coils and for the coils themselves to achieve their equilibrium configuration under compression may take many minutes or even days. This is one overriding feature that distinguishes interactions in polymer solutions from other interactions and one that makes theoretical modelling particularly difficult.

Chapter 14

Adhesion

14.1. Surface and interfacial energies

In Part I we saw how various interaction potentials between molecules arise, and we considered the implications of the energy minimum, that is, the "adhesion" energy of molecules in contact. Here we shall look at phenomena involving surfaces and particles in adhesive contact, and it is best to begin by defining some commonly used terms and deriving some important thermodynamic relations.

Work of adhesion and cohesion in vacuum. This is the free energy change, or reversible work done, to separate unit areas of two media 1 and 2 from contact to infinity in vacuum (Fig. 60a,b). For two different media ($1 \neq 2$), this energy is referred to as the *work of adhesion* W_{12}, while for two identical media ($1 = 2$), it becomes the *work of cohesion* W_{11}. If 1 is a solid and 2 a liquid, W_{12} is often denoted by W_{SL}. Note that since all media attract each other in vacuum W_{11} and W_{12} are always positive.

Surface energy and surface tension. This is the free energy change γ when the surface area of a medium is increased by unit area. Now the process of creating unit area of surface is equivalent to separating two half-unit areas from contact (Fig. 60b,c), so that we may write

$$\gamma_1 = \tfrac{1}{2} W_{11}. \tag{14.1}$$

For solids, γ_1 is commonly denoted by γ_S and is given in units of energy per unit area, millijoules per square meter (ergs per square centimeter). For liquids, γ_1 is commonly denoted by γ_L and is usually given in units of tension per unit length, millinewtons per meter (dynes per centimeter), which is numerically and dimensionally the same as the surface free energy.

213

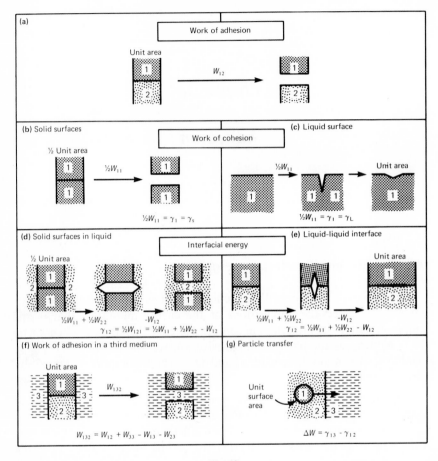

Fig. 60

It is evident that the intermolecular forces that determine the surface energy of a substance are the same as those that determine its latent heat and boiling point. As might be expected, substances such as metals with high boiling points ($T_B > 2000°C$) usually have high surface energies ($\gamma > 1000$ mJ m^{-2}), while lower boiling point substances have progressively lower surface energies [e.g., $\gamma = 485$ mJ m^{-2} for mercury ($T_B = 357°C$), $\gamma = 73$ mJ m^{-2} for water ($T_B = 100°C$), $\gamma = 13.2$ mJ m^{-2} for argon ($T_B = -186°C$), and $\gamma = 2.3$ mJ m^{-2} for hydrogen ($T_B = -253°C$)]. In Section 11.8 we saw how the surface energies of all but strongly polar or H-bonding liquids and solids can be calculated reasonably accurately on the basis of current theories of van der Waals forces.

It is important to appreciate that when the process of increasing the surface area of a medium takes place in a foreign vapour, such as laboratory air, some adsorption of vapour molecules (e.g., water, hydrocarbons) may take place on the newly created surface. This has the effect of lowering γ_S or γ_L from their real values and are denoted by γ_{SV} and γ_{LV} (Adamson, 1976, ch. 7). For example, when mica is cleaved in high vacuum $\gamma_S \approx 4500$ mJ m^{-2}, but when cleaved in laboratory air it falls to below 300 mJ m^{-2} (Bailey et al., 1970).

Interfacial energy.   When two immiscible liquids 1 and 2 are in contact, the free energy change in expanding their "interfacial" area by unit area is known as their *interfacial energy* or tension γ_{12}. The energetics associated with this expansion process may be understood by splitting it into two hypothetical steps (Fig. 60e): First, unit areas of media 1 and 2 are created, and are then brought into contact. The total free energy change γ_{12} is therefore

$$\gamma_{12} = \tfrac{1}{2}W_{11} + \tfrac{1}{2}W_{22} - W_{12} = \gamma_1 + \gamma_2 - W_{12}, \qquad (14.2)$$

which is often referred to as the *Dupré equation*. As shown in Fig. 60d, this energy is formally the same as that expended on separating two media 1 in medium 2 (W_{121}) or, conversely, of separating media 2 in medium 1 (W_{212}). We may therefore also write

$$\gamma_{12} = \tfrac{1}{2}W_{121} = \tfrac{1}{2}W_{212}. \qquad (14.3)$$

For a solid–liquid interface, γ_{12} is commonly denoted by γ_{SL}, so that the Dupré equation may be written as

$$\gamma_{SL} = \gamma_S + \gamma_L - W_{SL}. \qquad (14.4)$$

Table XIX gives the surface and interfacial energies of some common substances.

If only dispersion forces are responsible for the interaction between media 1 and 2, then we have previously seen that to a good approximation

$$W_{12}^d \approx \sqrt{W_{11}^d W_{22}^d} \approx 2\sqrt{\gamma_1^d \gamma_2^d},$$

so that Eq. (14.2) now becomes (Adamson, 1976, Chs. 3,7)

$$\gamma_{12} \approx \gamma_1 + \gamma_2 - 2\sqrt{\gamma_1^d \gamma_2^d}, \qquad (14.5)$$

where γ_1^d and γ_2^d are the dispersion force contributions to the surface tensions. Fowkes (1964) and Good and Elbing (1970) estimated that for water, the dispersion contribution to the total surface tension is 20 ± 2 mN m^{-1} or about 27% of the total,* the remaining 53 mN m^{-1} arising from nondispersion (i.e.,

---

* Note the near agreement between this value and the probably fortuitous theoretical estimates of 24 and 25% obtained in Tables X and XVII.

TABLE XIX
Surface and interfacial energies at 20°C

Liquid 1	Surface Energy γ_1 (mJ m^{-2})	Interfacial Energy $\gamma_{12}{}^a$ (mJ m^{-2})
		Against Water ($\gamma_2 = 72.75$)
Cyclohexanol	32.7	4
Diethyl ether	17.0	11
Chloroform	27.1	28
Benzene	28.8	35
Carbon tetrachloride	26.9	45
Cyclohexane	25.5	50
n-Hexane	18.4	51
n-Octane	21.8	51
n-Tetradecane	25.6	52
Paraffin wax	~25	48–52
		Against Glycol ($\gamma_2 = 47.7$)
Cyclohexane	25.5	14
n-Hexane	18.4	16

[a] Note the progressive increase in γ_{12} with increasing inertness (lower polarity, lower H-bonding capability) of liquid 1.

entropic, polar, and H-bonding) interactions. Since water and hydrocarbon attract each other mainly via dispersion forces the interfacial tension of a hydrocarbon–water interface should therefore be given by Eq. (14.5). Thus, for octane–water, putting $\gamma_1^d = 21.8$ mN m^{-1} and $\gamma_2^d = 20$ mN m^{-1}, we calculate

$$\gamma_{12} \approx 21.8 + 72.75 - 2\sqrt{21.8 \times 20} \approx 52.8 \quad \text{mN m}^{-1},$$

which is very close to the measured value of 50.8 mN m^{-1}. This good agreement is obtained for many hydrocarbon–water interfaces, but the agreement is not so good for aromatic molecules such as benzene and toluene (Fowkes, 1964; Good and Elbing, 1970).

Work of adhesion in a third medium. It is left as an exercise for the reader to establish that the energy change on separating two media 1 and 2 in medium 3 (Fig. 60f) is given by

$$W_{132} = W_{12} + W_{33} - W_{13} - W_{23} = \gamma_{13} + \gamma_{23} - \gamma_{12}. \qquad (14.6)$$

If medium 3 is vacuum, $W_{132} \to W_{12}$, $\gamma_{13} \to \gamma_1$, $\gamma_{23} \to \gamma_2$, and the above reduces to Eq. (14.2) as expected.

Surface energy of transfer.   When a macroscopic particle 1 moves from medium 2 to medium 3 (Fig. 60g) the change in energy per unit area of the particle's surface is

$$\Delta W = (W_{12} - \tfrac{1}{2}W_{22}) - (W_{13} - \tfrac{1}{2}W_{33}) = \gamma_{13} - \gamma_{12}, \qquad (14.7)$$

where $(W_{12} - \tfrac{1}{2}W_{22})$ is the energy required to first separate unit areas of media 1 and 2 and then bring into contact the free surfaces of medium 2 and $(W_{13} - \tfrac{1}{2}W_{33})$ is the reverse operation with medium 3.

In Part III we shall use this expression when considering the association of amphiphilic molecules into micelles and bilayers. Here we shall briefly investigate the validity of applying surface and interfacial energies to very small droplets and molecules. Clearly the idea of a surface *tension* cannot apply to a single molecule or even to a cluster of a few molecules. But the concept of a surface *energy* remains valid, even for isolated molecules since there is always an energy change associated with transferring a molecule from one medium to another, and this can always be expressed in terms of $4\pi a^2(\gamma_{13} - \gamma_{12})$, where a is the molecular radius. The question is, How close will the surface energy of a small cluster correspond to that of the planar interface? There is at present no simple answer to this problem (Sinanoğlu, 1981), but we may gain a feel for the size limit by comparing the nearest-neighbour interactions between simple spherical molecules at a planar surface with those occurring at a highly curved surface. If $-w$ is the pair energy at molecular contact, then for a planar close packed surface lattice, with three unsaturated bonds per surface molecule, the surface energy is given by Eq. (11.28):

$$\gamma \approx 3w/(2\sigma^2 \sin 60°) \approx 1.7w/\sigma^2.$$

For an isolated molecule, with 12 unsaturated bonds, its effective surface energy will be

$$\gamma = 12w/2[4\pi(\sigma/2)^2] \approx 1.9w/\sigma^2, \qquad (14.8)$$

while for a cluster of 13 molecules, with seven unsaturated bonds per each of the 12 surface molecules (Fig. 61), we find

$$\gamma \approx 12 \times 7w/2[4\pi(3\sigma/2)^2] \approx 1.5w/\sigma^2. \qquad (14.9)$$

By symmetry, it is also clear that cavities of one or 13 (missing) molecules must also have the same surface energies as above. Thus, we arrive at the remarkable conclusion that the magnitude of the effective surface energy γ of a very small cluster or even an isolated molecule is similar to that of a planar macroscopic surface (Sinanoğlu, 1981). We encounter further experimental manifestations of this phenomenon in Sections 8.5 and 14.4 and again in Part III.

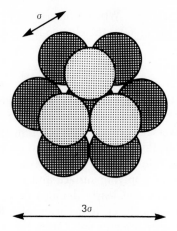

Fig. 61 Cluster of 13 molecules: one central molecule surrounded by 12 close-packed neighbours (six in the plane of the page with three above and three below). Note that each of the 12 surrounding molecules has five contact points with its neighbours.

14.2. Contact angles

A consideration of surface and interfacial energies is also important for determining how liquid droplets deform as they approach a surface. In Fig. 62a (top) a large initially spherical droplet 2 in medium 3 approaches and then settles on the rigid flat surface of medium 1. The final total surface energy of the system is therefore

$$W_{\text{tot}} = \gamma_{23}(A_{\text{c}} + A_{\text{f}}) - W_{132}A_{\text{f}}, \tag{14.10}$$

where A_{c} and A_{f} are the curved and flat areas of the droplet. At equilibrium: $\gamma_{23}(dA_{\text{c}} + dA_{\text{f}}) - W_{132}\, dA_{\text{f}} = 0$. For a droplet of constant volume, it is easy to show using straightforward geometry that $dA_{\text{c}}/dA_{\text{f}} = \cos\theta$. Thus, the equilibrium condition ($\theta = \theta_0$) is

$$\gamma_{23}(1 + \cos\theta_0) = W_{132} = \gamma_{13} + \gamma_{23} - \gamma_{12}, \tag{14.11}$$

or

$$\gamma_{12} + \gamma_{23}\cos\theta_0 = \gamma_{13}, \tag{14.12}$$

which is often derived by balancing the resolved interfacial tensions in the plane of the surface. If media 2 and 3 are interchanged, Fig. 62a (bottom), then Eq. (14.11) becomes

$$\gamma_{23}(1 + \cos\theta_0) = W_{123} = \gamma_{12} + \gamma_{23} - \gamma_{13}, \tag{14.13}$$

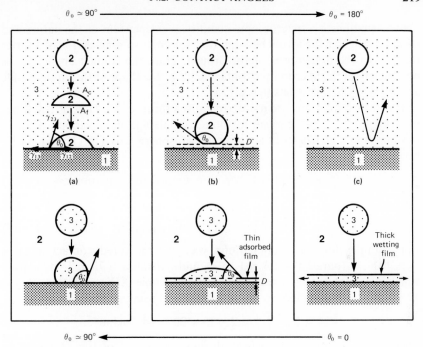

Fig. 62 Contact angles can often be changed by chemically modifying surfaces or by addition of certain solute molecules into the medium that adsorb on the surfaces. For example, addition of "surface-active" molecules such as detergents to water can cause an oil-surface contact angle to increase from 0 to 180° (see Fig. 2d). When quartz is preheated above 300°C its hydrophilic surface silanol groups —Si(OH)—Si(OH)— give off water, leaving behind hydrophobic siloxane groups —Si—O—Si—, and the contact angle rises from 0 to about 60°. Note that the upper and lower drawings in each box are formally equivalent [with media 2 and 3 interchanged so that $\theta_0(\text{bottom}) = 180° - \theta_0(\text{top})$].

that is,

$$\gamma_{12} - \gamma_{23} \cos \theta_0 = \gamma_{13}, \qquad (14.14)$$

or

$$\gamma_{23}(1 - \cos \theta_0) = W_{132}, \qquad (14.15)$$

which is the same as Eq. (14.11) with θ_0 replaced by $180° - \theta_0$. Thus, as might have been expected intuitively, the contact angle in Fig. 62a (bottom) is simply $180° - \theta_0$ of that in Fig. 62a (top).

While the above results were derived for the case of a droplet on a flat surface, the angle subtended by a liquid on a solid remains the same irrespective of the surface geometry (Adamson, 1976, Ch. 7), for example, on curved surfaces or in capillaries. Further, the contact angle θ_0 as given by the above equations is a macroscopic quantity—independent of the nature and effective range of the forces between the molecules so long as it is of shorter range than the dimensions of the droplet. Thus, the contact angle tells us nothing about the microscopic contact angle or the shape of the liquid profile at the point where it meets the surface.

The above equations are more generalized forms of the famous equations of Young and Dupré derived for liquid droplets in inert vapour. Thus, if medium 3 in Fig. 62a (top) is an inert atmosphere, Eqs. (14.11) and (14.12) become

$$\gamma_2(1 + \cos\theta_0) = W_{12} \qquad \text{(Young–Dupré equation)} \qquad (14.16)$$

and

$$\gamma_{12} + \gamma_2 \cos\theta_0 = \gamma_1 \qquad \text{(Young equation).} \qquad (14.17)$$

For example, for water on paraffin wax, the measured contact angle is $\theta_0 \approx 111°$, $\gamma_1(\text{paraffin}) \approx 25$ mJ m^{-2}, and $\gamma_2(\text{water}) = 73$ mJ m^{-2}, from which we infer that $\gamma_{12} \approx 51$ mJ m^{-2} (cf. Table XIX) and that $W_{12} \approx 47$ mJ m^{-2}, which is close to the value expected using

$$W_{12} \approx 2\sqrt{\gamma_1^d \gamma_2^d} \approx 2\sqrt{25 \times 20} \approx 45 \quad \text{mJ m}^{-2}.$$

There is one reported case where all four parameters of the Young equation have been independently measured (Pashley and Israelachvili, 1981; Israelachvili, 1982). This concerns a droplet of 8×10^{-4} M HTAB solution on a monolayer-covered surface of mica for which $\gamma_1(\text{solid}) = 27 \pm 2$ mJ m^{-2}, $\gamma_2(\text{liquid}) = 40$ mJ m^{-2}, $\gamma_{12}(\text{solid–liquid}) = 11 \pm 2$ mJ m^{-2}, and $\theta_0 = 64°$, which agree with the Young equation.

In the case of medium 3 being a liquid, equilibrium can be attained at some finite distance D [Fig. 62b (top)] where the interaction energy W_{132} is a minimum (e.g., a weak secondary minimum) (Section 12.17). In such cases the contact angle is usually very low. Such phenomena occur, for example, when dissolved air bubbles or oil droplets containing lipid or surfactant monolayers (emulsion droplets) adhere weakly to each other or to a surface (Fig. 2d).

An analogous situation occurs when a liquid droplet surrounded by its vapour attaches to a surface containing a thin physisorbed film of the liquid [Fig. 62b (bottom)]. Qualitatively one may say that here a small contact angle forms because the liquid rests on a surface that is of its own kind. Note that this is formally the same as that of Fig. 62b (top) with media 2 and 3 interchanged. Such cases occur quite often, for example, many different vapours including water and hydrocarbons adsorb as a monolayer on mica, and the liquids have a small but finite contact angle on mica of $\theta_0 \lesssim 6°$.

Finally we may note that if the interaction between 1 and 2 across 3 is monotonically repulsive, the liquid droplet is now repelled from the surface [Fig. 62c (top)], and we must put $W_{132} = 0$. Such situations lead to the complete spreading of a liquid on a surface and the development of thick wetting films [Fig. 62c (bottom)], previously discussed in Sections 11.5 and 12.8.

14.3. Adhesion force between solid particles

The adhesion force of two rigid (incompressible) macroscopic spheres is simply related to their work of adhesion by

$$F = 2\pi\left(\frac{R_1 R_2}{R_1 + R_2}\right)W_{132}. \tag{14.18}$$

This general result is a direct consequence of the Derjaguin approximation, Eq. (10.16), and leads to the following special cases:

$$F = 2\pi R\gamma_{SL} \quad \text{(two identical spheres in liquid),} \tag{14.18a}$$

$$F = 2\pi R\gamma_{S} \quad \text{(two identical spheres in vacuum),} \tag{14.18b}$$

$$F = 4\pi R\gamma_{S} \quad \text{(sphere on surface in vacuum),} \tag{14.18c}$$

$$F = 4\pi R\gamma_{SV} \quad \text{(sphere on surface in vapour).} \tag{14.18d}$$

Real particles, however, are never completely rigid, and on coming into contact they deform elastically under the influence of the attractive intersurface forces that pull the two surfaces together giving rise to a finite contact radius a_0 even under zero external load (Fig. 63). One of the first attempts at a

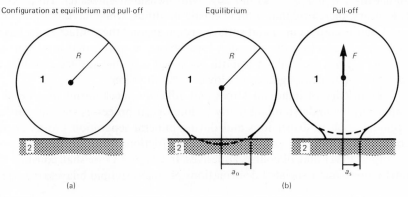

Fig. 63 (a) Rigid sphere on rigid surface. (b) Deformable (elastic) sphere on rigid surface. Rigid and elastic bodies that adhere to each other undergo very different deformations on coming into contact and on being separated. Elastic bodies separate spontaneously when their contact area is still finite, but the adhesion force is essentially the same $F \approx 2\pi R W_{12}$ for both rigid and elastic bodies, so long as the deformation is small. For two spheres (instead of a sphere on a plane), $F \approx \pi R W_{12}$.

rigorous theoretical treatment of the adhesion of elastic spheres is due to Johnson *et al.* (1971), whose theory, the *JKR theory*, predicts that on pulling two elastic spheres from adhesive contact the contact radius decreases gradually from a_0 to $a_s = (0.25)^{1/3} a_0 = 0.63 a_0$ at which point the spheres spontaneously separate. The predicted adhesion force, or "pull-off" force, is 75% of that given by Eq. (14.18) for undeformable rigid spheres (a value that is, surprisingly, independent of the original area of contact a_0 and the elastic modulus of the spheres). Advances in theories of adhesion have been made by Derjaguin *et al.* (1975), Tabor (1977), Muller *et al.* (1980, 1983), and Savkoor (1982).

Israelachvili *et al.* (1980b) measured the adhesion force in an inert atmosphere between two curved ($R \approx 1$ cm) molecularly smooth hydrocarbon surfaces of known surface energies, $\gamma_s \approx 25$ mJ m^{-2}, and found that

(i) separation occurs spontaneously once the contact radius has fallen to 0.58 ± 0.04 of the equilibrium radius a_0, in good agreement with the JKR theory, and

(ii) the adhesion force is given by Eq. (14.18c) (i.e., about the value expected for rigid bodies and 25% higher than expected from the JKR theory).

Shchukin and co-workers (Shchukin *et al.*, 1981; Shchukin, 1982) measured the adhesion forces between smooth methylated (hydrophobic) glass spheres in air and in various liquids. By use of Eq. (14.18a,b) they deduced the following values for the surface and interfacial energies: $\gamma_S \approx 20$ mJ m^{-2}, γ_{SL}(hydrocarbon–water) ≈ 40 mJ m^{-2}, and γ_{SL}(hydrocarbon–glycol) ≈ 16 mJ m^{-2}, which agree within 20% with measured values (cf. Table XIX). It appears, therefore, that a finite elastic modulus, while having a dramatic effect on the contact area and surface profile around the contact zone, has at most a 25% effect on the adhesion force. Note, however, that most particle surfaces are not molecularly smooth but rough and that surface esperities as small as 1–2 nm can significantly lower their adhesion.

A knowledge of adhesion forces and adhesion mechanisms in general is important for understanding many technological processes such as particle agglomeration in dispersions and during mineral separation processes, the cohesive strength of powders and soils, friction, lubrication, and wear (Georges, 1982), and, of course, adhesives. In Chapter 17 we shall consider the interactions and associated deformations of spherical lipid bilayers (vesicles).

14.4. Effect of capillary condensation on adhesion

The mechanical and adhesive properties of many substances are very sensitive to the presence of even trace amounts of vapours in the atmosphere. For example, the adhesion of powders (Visser, 1976), the strength of quartz

(Paterson and Kekulawala, 1979), and the seismic properties of rocks (Clark *et al.*, 1980) are markedly dependent on the relative humidity. All these effects are due in part to the *capillary condensation* of water around surface contact sites (e.g., in cracks and pores), which, as we shall see, can have a profound effect on the strength of adhesion joints.

Liquids that wet or have a small contact angle on surfaces will condense from vapour into cracks and pores as bulk liquid (Fig. 64a,b). At equilibrium the meniscus curvature $(1/r_1 + 1/r_2)$ is related to the relative vapour pressure (relative humidity for water) p/p_s by the *Kelvin equation* (Adamson, 1976, Ch. 2)

$$\left(\frac{1}{r_1} + \frac{1}{r_2}\right)^{-1} = r_K = \frac{\gamma V}{RT \log(p/p_s)}, \qquad (14.19)$$

where r_K is the *Kelvin radius* and V is the molar volume ($\gamma V/RT = 0.54$ nm for water at $20°C$). Thus, for a spherical concave water meniscus ($r_1 = r_2 = r$),

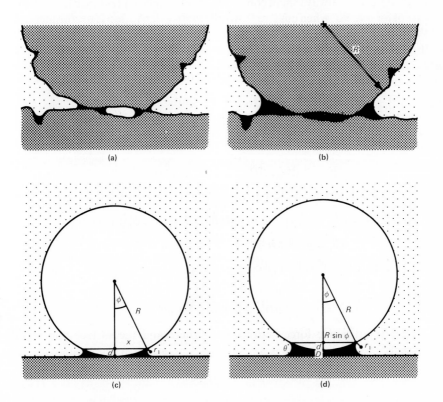

(a)   (b)

(c)   (d)

Fig. 64

we find $r = \infty$ at $p/p_s = 1$, $r \approx -10$ nm at $p/p_s = 0.9$, $r \approx -1.6$ nm at $p/p_s = 0.5$, and $r \approx -0.5$ nm at $p/p_s = 0.1$.

What is the effect of a liquid condensate on the adhesion force between a macroscopic sphere and a surface (Fig. 64c)? A simple derivation is to consider the *Laplace pressure* in the liquid (Adamson, 1976, Ch. 1):

$$P = \gamma_L \left(\frac{1}{r_1} + \frac{1}{r_2} \right) \approx \frac{\gamma_L}{r_1} \qquad \text{(since } r_2 \gg r_1 \text{).} \qquad (14.20)$$

The Laplace pressure acts on an area $\pi x^2 \approx 2\pi R d$ between the two surfaces thus pulling them together with a force

$$F \approx (\gamma_L/r_1)2\pi R d,$$

and since for small ϕ, $d \approx 2r_1 \cos \theta$ obtain

$$F \approx 4\pi R \gamma_L \cos \theta. \qquad (14.21)$$

The additional force arising from the resolved surface tension around the circumference is always small compared to the Laplace pressure contribution, except for $\theta \approx 90°$ when $\cos \theta \approx 0$. An alternative derivation is to consider how the total surface free energy of the system W_{tot} changes with separation D (Fig. 64d). For small ϕ,

$$-W_{tot} \approx 2\pi R^2 \sin^2 \phi (\gamma_{SL} - \gamma_{SV}) + \text{constant} + \text{smaller terms}$$

$$\approx 2\pi R^2 \phi^2 (\gamma_{SL} - \gamma_{SV}) + \text{constant}$$

$$= 2\pi R^2 \phi^2 \gamma_L \cos \theta + \text{constant},$$

so that

$$F = -dW_{tot}/dD = +4\pi R^2 \phi \gamma_L \cos \theta \, d\phi/dD.$$

Now if the liquid volume V remains constant, then since

$$V \approx \pi R^2 \sin^2 \phi (D + d) - (\pi R^3/3)(1 - \cos \phi)^2 (2 + \cos \phi)$$

$$\approx \pi R^2 D \phi^2 + \pi R^3 \phi^4/3 \qquad \text{for small } \phi$$

we have $dV/dD = 0$, which gives

$$\frac{d\phi}{dD} = -\frac{1}{(R\phi + 2D/\phi)}.$$

Thus, the attractive force between the sphere and the surface due to the presence of a liquid bridge is

$$F = \frac{4\pi R \gamma_L \cos \theta}{(1 + D/d)}, \qquad (14.22)$$

and at $D = 0$

$$F = F_{max} = 4\pi R\gamma_L \cos\theta, \qquad (14.23)$$

which is the same as Eq. (14.21). More rigorous expressions, valid for large ϕ and different contact angles on each surface, are given by Orr $et\ al.$ (1975). One other important parameter must be included in the above expression. This is the solid–solid contact adhesion force inside the liquid annulus, Eq. (14.18). The final result for a small meniscus of radius r_1 (strictly $R \gg r_1$) is therefore

$$F = 4\pi R[\gamma_L \cos\theta + \gamma_{SL}] = 4\pi R\gamma_{SV}, \qquad (14.24)$$

which may be compared with

$$F = 4\pi R\gamma_S \qquad (14.25)$$

in the absence·of any condensing vapour (i.e., as $r_1 \to 0$). For two spheres, R should be replaced by $(1/R_1 + 1/R_2)^{-1}$.

Since $\gamma_S > \gamma_{SV}$ the adhesion force should always be *less* in a vapour than in vacuum. In practice, however, one often has to compare γ_{SV_1} with γ_{SV_2} depending on the relative adsorptions of different vapours from the atmosphere, and the adhesion force in air may *increase* with relative humidity if γ_{SV_1} (moist air) $> \gamma_{SV_2}$ (dry air). Equation (14.24) shows that the adhesion force in vapour must exceed $4\pi R\gamma_L \cos\theta$. Often $\gamma_L \cos\theta$ greatly exceeds γ_{SL}, whence the adhesion force is determined solely by the surface energy of the liquid (e.g., water) given by Eq. (14.23).

Note that Eqs. (14.24) and (14.25) are independent of the meniscus radius r_1 so that it is of interest to establish below what radius, or relative vapour pressure, these equations do break down. McFarlane and Tabor (1950) verified that the adhesion force between glass spheres in saturated vapours of water, glycerol, decane, octane, alcohol, benzene, and aniline are all given by $F = 4\pi R\gamma_L \cos\theta$ to within a few percentage points. Fisher and Israelachvili (1981) measured the adhesion forces between curved mica surfaces in various vapours such as cyclohexane and benzene and found that $F = 4\pi R\gamma_L \cos\theta$ is already valid once the relative vapour pressures exceed 0.1–0.2, corresponding to meniscus radii of only ~ 0.5 nm (i.e., about the size of the molecules). However, for water, a larger radius of ~ 5 nm was needed before Eq. (14.23) was satisfied. These results confirm the suggestion made in Section 14.1 that for molecules that interact mainly via a simple Lennard-Jones potential, their bulk surface energy is already manifest at very small curvatures.

Since real particle surfaces are often rough their adhesion in vapour is not always given by Eq. (14.24). For example, the adhesion of dry sand particles is very small, and even when slightly moist the adhesion is not much different since the condensed water is only bridging small asperities (Fig. 64a). However, once r_K exceeds the asperity size but is still less than the particle radius R

(Fig. 64b), the adhesion force attains its full value of $F \approx 4\pi R \gamma_{\text{L}} \cos \theta$ (McFarlane and Tabor, 1950). Finally, when sand is completely wet the adhesion will once again be very low. It is for this reason that one can only build sandcastles with moist sand, but not with dry or completely wet sand (R. Pashley, unpublished results).

Capillary condensation also occurs when water condenses from a solvent in which it is only sparingly soluble (e.g., from hydrocarbon solvents where the solubility is usually below 100 ppm). In such circumstances the presence of even 20 ppm (i.e., 20–40% of saturation) can lead to a dramatic increase in the adhesion of colloidal particles. Indeed, it has long been known that trace amounts of water can have a dramatic effect on colloidal stability (Bloomquist and Shutt, 1940; Parfitt and Peacock, 1978) and surfactant association (Eicke, 1980) in nonpolar organic solvents; and the enhanced agglomeration of metal ores and coal particles in oils by addition of water forms the basis of several industrial separation (extraction) processes (Henry et al., 1980). Christenson (1983) found that in the presence of small amounts of water (< 100 ppm) the adhesion force between mica surfaces in benzene, octane, cyclohexane, and OMCTS is given by Eq. (14.23) to within 20% with γ_{L} replaced by the liquid–liquid interfacial energy γ_{12}. The enhanced adhesion in such systems arises because γ_{12} is typically 35–50 mJ m^{-2} (Table XIX)—much higher than the solid–liquid interfacial energies in the anhydrous liquids.

PART THREE

FLUID-LIKE STRUCTURES: MICELLES, BILAYERS, AND BIOLOGICAL MEMBRANES

Chapter 15

Thermodynamic Principles
of Self-Association

15.1. Introduction

In Part III we shall be looking at the interactions of small molecular aggregates such as micelles, bilayers, vesicles, and biological membranes, which form readily in aqueous solution by the spontaneous self-association of certain amphiphilic molecules (see Figs. 65 and 66 and Table XX). These structures—sometimes collectively referred to as *association colloids*—stand apart from the conventional colloidal particles discussed in Part II in one important respect: Unlike solid particles or rigid macromolecules such as DNA, they are soft and flexible (i.e., *fluidlike*). This is because the forces that hold amphiphilic molecules together in micelles and bilayers are not due to strong covalent or ionic bonds but arise from weaker van der Waals, hydrophobic, hydrogen-bonding, and screened electrostatic interactions. Thus, if the solution conditions, such as the electrolyte concentration or the pH, of an aqueous suspension of micelles or vesicles is changed, not only will this affect the interactions between the aggregates but it will also affect the forces between the molecules *within* each aggregate, thereby modifying the size and shape of the structures themselves. It is therefore necessary to begin by considering the factors that determine how and why certain molecules associate into various well-defined structures.

In Chapters 15 and 16 we shall be concerned with the thermodynamic and physical principles of self-association in general and of amphiphilic molecules such as surfactants and lipids in particular, while in the final Chapter 17 we shall investigate the van der Waals forces, electrostatic (double-layer) forces, and the solvation (hydration) forces between bilayers and membranes, and we

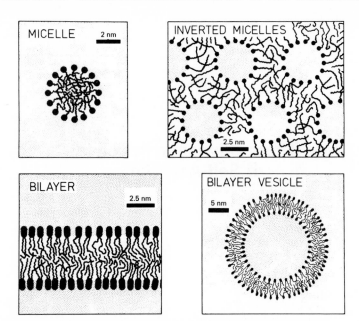

Fig. 65 Amphiphiles such as surfactants and lipids (see Table XX) can associate into a variety of structures in aqueous solutions, which can transform from one to another by changing the solution conditions, such as the electrolyte or lipid concentration, pH, or temperature. In most cases the hydrocarbon chains are in the fluid state allowing for the passage of water and ions across the hydrophobic regions of bilayers. The lifetime of water molecules in lecithin vesicles is about 0.02 sec, while ions can be trapped for much longer times, about 8 hr for Cl^- and one month for Na^+ ions. Most single-chained surfactants form micelles, while most double-chained surfactants form bilayers, for reasons that are discussed in Chapter 16.

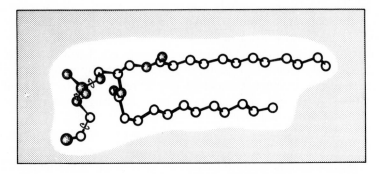

Fig. 66 The zwitterionic phospholipid dilauryl-phosphatidyl-ethanolamine containing two saturated hydrocarbon chains and a polar (hydrophilic) head group (see also Table XX).

shall examine their role in the interactions of, for example, lipid vesicles and biological membranes.

Our first concern will be to formulate the basic equations of self-association in general statistical thermodynamic terms and then go on to investigate the relevant intermolecular interactions that determine into which structures different amphiphiles will assemble. We shall find that a very beautiful picture emerges that brings out the role of molecular geometry in determining the structures formed, and from which many of the physical properties of these structures can be quantitatively understood without requiring a detailed knowledge of the very complex short-range forces operating between the polar head groups and the hydrocarbon chains. [By analogy, the van der Waals equation of state contains no information on the nature and range of intermolecular forces (i.e., the force laws) and yet provides a very satisfactory description of gas–liquid phase behaviour. Indeed, when van der Waals in 1873 proposed his famous equation he knew nothing about the origin and nature of van der Waals forces.]

15.2. Fundamental thermodynamic equations of self-association

The literature on this subject is voluminous and often confusing, the most rigorous treatment being that of Hall and Pethica (1967) based on Hill's classic book on small systems thermodynamics (Hill, 1963, 1964). We shall follow the more simplified approach and notation of Tanford (1973, 1980) for micelles, which was later extended to larger lipid aggregates such as vesicles and bilayers by Israelachvili et al. (1976, 1977) and recently reviewed by Wennerström and Lindman (1979) and Israelachvili et al. (1980a).

Equilibrium thermodynamics requires that in a system of molecules that form aggregated structures in solution (Fig. 67) the chemical potential of all identical molecules in different aggregates be the same. This may be expressed as

$$\mu = \mu_1^0 + kT \log X_1 = \mu_2^0 + \tfrac{1}{2}kT \log \tfrac{1}{2}X_2 = \mu_3^0 + \tfrac{1}{3}kT \log \tfrac{1}{3}X_3 = \cdots$$
$$\text{monomers} \qquad\qquad \text{dimers} \qquad\qquad \text{trimers}$$

or

$$\mu = \mu_N = \mu_N^0 + \frac{kT}{N} \log\left(\frac{X_N}{N}\right) = \text{constant}, \qquad N = 1, 2, 3, \ldots, \quad (15.1)$$

where μ_N is the chemical potential of a molecule in an aggregate of aggregation number N (the same for all N), μ_N^0 the standard part of the chemical potential (the mean interaction free energy) *per molecule* in aggregates of aggregation

TABLE XX
Some common amphiphiles

Single-chained Surfactants

Anionic	$C_{12}H_{25}-O-SO_3^- Na^+$	Sodium dodecyl sulphate (SDS or NaDS)
Anionic	$C_{18}H_{37}-COO^- H^+$	Stearic acid
Cationic	$C_{16}H_{33}-N^+(CH_3)_3 Br^-$	Hexadecyl trimethylammonium bromide (HTAB or CTAB)
Nonionic	$C_{12}H_{25}-(O-CH_2-CH_2)_4OH$	Tetraoxyethylene dodecylether ($C_{12}E_4$)
Zwitterionic	Single chained lecithin (see below)	Lysolecithin

Double-chained Phospholipids

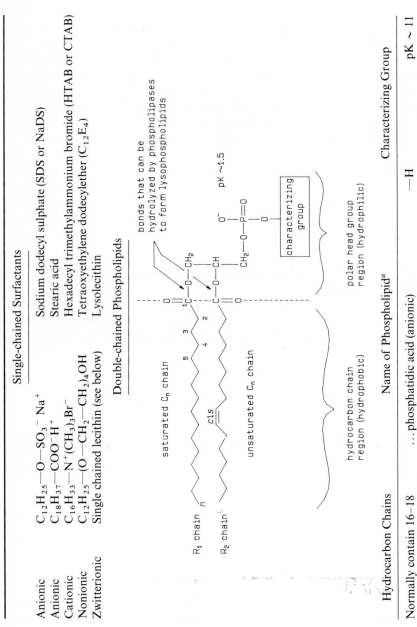

Hydrocarbon Chains	Name of Phospholipid[a]	Characterizing Group
Normally contain 16–18 carbons per chain, the R_2 chain containing 1–3 cis	...phosphatidic acid (anionic)	—H pK ~ 11

232

diC$_{12}$: dilauroyl... ...phosphatidyl choline or lecithin (zwitterionic) $-CH_2-CH_2-N^+(CH_3)_3$

diC$_{14}$: dimyristoyl... ...phosphatidyl ethanolamine (zwitterionic) $-CH_2-CH_2-NH_3^+$ pk $\sim$ 11

diC$_{16}$: dipalmitoyl... ...phosphatidyl glycerol (anionic) $-CH_2-CH$ $\begin{matrix} CH_2OH \\ OH \end{matrix}$

diC$_{18}$: distearoyl... ...phosphatidyl serine (anionic) $-CH_2-CH$ $\begin{matrix} COO^- \\ NH_3^+ \end{matrix}$

Other Double-chained Surfactants and Lipids[b]

Dihexadecyl dimethylammonium bromide (diC$_{16}$DAB)

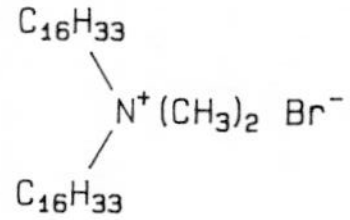

$C_{16}H_{33}$
$N^+(CH_3)_2$ Br$^-$
$C_{16}H_{33}$

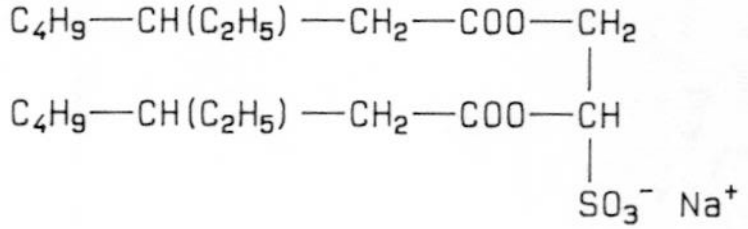

$C_4H_9-CH(C_2H_5)-CH_2-COO-CH_2$
$C_4H_9-CH(C_2H_5)-CH_2-COO-CH$
SO_3^- Na$^+$

Aerosol OT

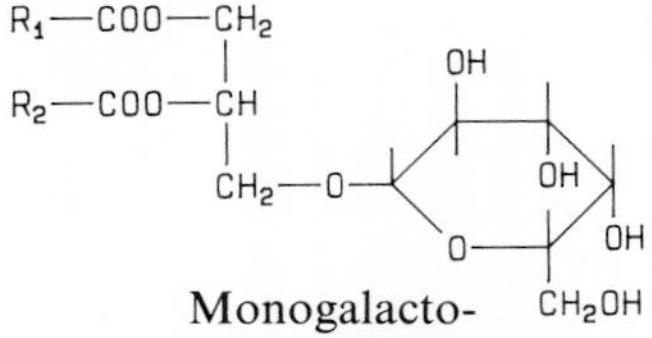

Monogalacto- syldiglyceride (MGDG)

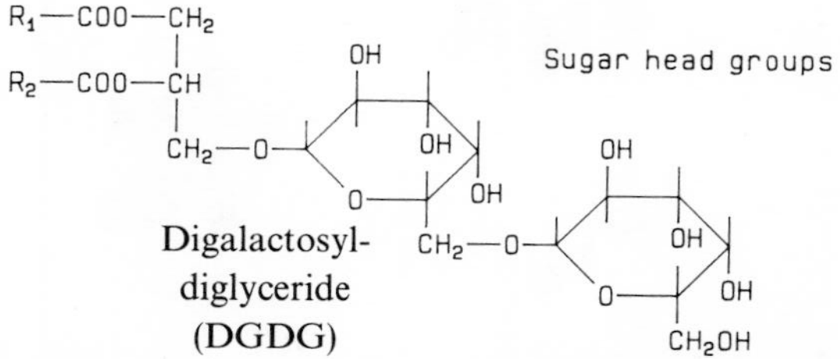

$R_1-COO-CH_2$
$R_2-COO-CH$
CH_2-O

Digalactosyl- diglyceride (DGDG)

$R_1-COO-CH_2$
$R_2-COO-CH$
CH_2-O

Sugar head groups

---

[a] The ionic states of the head groups are given for aqueous dispersions at pH 7. At high pH (>11.5) phosphatidylethanolamine becomes negatively charged while at low pH (<1) it becomes positively charged.
[b] Phosphatidylcholines and phosphatidylethanolamines are the two major lipids found in animal membranes, while the galactolipids DGDG and MGDG are the major constituents of plant thylakoid membranes. Note that none of these carry a net charge at normal pH.

233

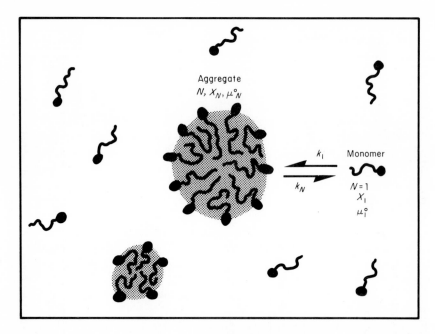

Fig. 67 Association of N monomers into a aggregate (e.g., a micelle). The mean lifetime of an amphiphilic molecule in a micelle is very short, typically $10^{-5}-10^{-3}$ s.

number N, and X_N the concentration (more strictly the activity) of molecules incorporated in aggregates of number N ($N = 1$, μ_1^0 and X_1 correspond to isolated molecules, or *monomers*, in solution). Equation (15.1) may also be derived using the familiar *law of mass action* (Alexander and Johnson, 1950). Thus, referring to Fig. 67 we may write

$$\text{rate of association} = k_1 X_1^N,$$

$$\text{rate of dissociation} = k_N(X_N/N),$$

where

$$k_1/k_N = \exp[-N(\mu_N^0 - \mu_1^0)/kT]$$

is the ratio of the two "reaction" rates (the equilibrium constant). These combine to give Eq. (15.1), which can also be written in the more useful (and equivalent) forms

$$X_N = N\{X_1 \exp[(\mu_1^0 - \mu_N^0)/kT]\}^N \tag{15.2}$$

and

$$X_N = N\{(X_M/M) \exp[M(\mu_M^0 - \mu_N^0)/kT]\}^{N/M}, \tag{15.3}$$

where M is any arbitrary reference state of aggregates (or monomers) with aggregation number M (or 1). Equation (15.2) together with the conservation relation relating total solute concentration C as

$$C = X_1 + X_2 + X_3 + \cdots = \sum_{N=1}^{\infty} X_N \qquad (15.4)$$

completely defines the system. Depending on how the free energies μ_1^0, μ_N^0 are defined the dimensionless concentrations C and X_N can be expressed as volume fraction or mole fraction [(mol dm^{-3})/55.5 for aqueous solutions]. In particular, note that C and X_N can never exceed unity. Equation (15.2) assumes ideal mixing (i.e., it is restricted to dilute systems where interaggregate interactions can be ignored). The effect of such interactions will be considered later.

Little more can be said about aggregated dispersions without specifying the form and magnitude of μ_N^0 as a function of N. This important matter will now be considered, and it is instructive to first proceed with a formal thermodynamic analysis of the above equations.

15.3. Conditions necessary for the formation of aggregates

Aggregates form only when there is a difference in the cohesive or binding energies between the molecules in the aggregated and the dispersed (monomer) states. If the molecules in different sized aggregates (including monomers) all experience the *same* interaction with their surroundings, the value of μ_N^0 will remain constant in different aggregates (with different N), and Eq. (15.2) becomes

$$X_N = NX_1^N \qquad \text{for} \quad \mu_1^0 = \mu_2^0 = \mu_3^0 = \cdots = \mu_N^0. \qquad (15.5)$$

And since $X_1 < 1$, we must have $X_N \ll X_1$ so that most of the molecules will be in the monomer state ($N = 1$). If μ_N^0 increases as N increases, Eq. (15.2) shows that the occurrence of large aggregates becomes even less probable.

The necessary condition for the formation of large stable aggregates is that $\mu_N^0 < \mu_1^0$ for some value of N (e.g., when μ_N^0 progressively *decreases* as N increases or when μ_N^0 has a minimum value at some finite value of N). As we shall see, the exact functional variation of μ_N^0 with N also determines many of the physical properties of aggregates (e.g., their mean size and polydispersity). Since this variation may be a complex one it is clear that a number of structurally different populations may coexist and yet be in thermodynamic equilibrium with each other [note that X_N in Eq. (15.2) is a *distribution function* and may peak at more than one value of N].

We shall now consider the functional forms of μ_N^0 for some simple structures and by use of Eqs. (15.2)–(15.4) investigate their physical properties.

15.4. Variation of μ_N^0 with N for simple structures of different geometries: rods, discs, and spheres

One-dimensional aggregates (*rods*). As mentioned above, aggregates will form if μ_N^0 decreases with N. We shall now see that the dependence of μ_N^0 on N is usually determined by the geometrical shape of the aggregate. Let us begin by considering a suspension of rod-like aggregates made up of linear chains of identical molecules or monomer units in equilibrium with monomers in solution, where αkT is the free energy of binding a monomer to the aggregate in the solvent (i.e., the monomer–monomer "bond" energy in the aggregate) (Fig. 68). The total interaction free energy $N\mu_N^0$ of an aggregate of N monomers is therefore (remembering that the terminal monomers are unbonded)

$$N\mu_N^0 = -(N - 1)\alpha kT,$$

that is,

$$\mu_N^0 = -(1 - 1/N)\alpha kT = \mu_\infty^0 + \alpha kT/N. \tag{15.6}$$

Thus, as N increases the mean free energy μ_N^0 decreases asymptotically towards μ_∞^0, the "bulk" energy of a molecule in an infinite aggregate. A similar expression for μ_N^0 is obtained for any type of rod-like structure (e.g., a cylindrical micelle).

Two-dimensional aggregates (*discs*). Let us now look at disc-like or sheet-like aggregates (Fig. 68). Here the number N of molecules per disc is proportional to the area πR^2, while the number of unbonded molecules in the rim is proportional to the circumference $2\pi R$, and hence to $N^{1/2}$. The mean free energy per molecule in such an aggregate is therefore

$$\mu_N^0 = \mu_\infty^0 + \alpha kT/N^{1/2}, \tag{15.7}$$

where again α is a constant characteristic of the monomer–monomer and monomer–solvent interaction.

Three-dimensional aggregates (*spheres*). Finally, let us consider spherical aggregates or small solute droplets of radius R in a solvent (Fig. 68). Here N is proportional to the volume $\frac{4}{3}\pi R^3$, while the number of unbonded surface molecules is proportional to the area $4\pi R^2$ and hence to $N^{2/3}$. We therefore have

$$\mu_N^0 = \mu_\infty^0 + \alpha kT/N^{1/3}. \tag{15.8}$$

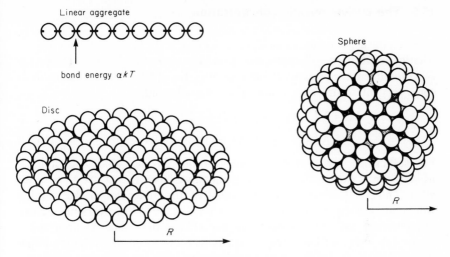

Linear aggregate

bond energy αkT

Sphere

Disc

R

R

Fig. 68 One-, two-, and three-dimensional structures formed by the association of identical monomer units in solution.

As an example, consider the association of small hydrocarbon molecules such as alkanes in water. If v is the molecular volume per molecule, then $N = 4\pi R^3/3v$. The free energy of the sphere is given by $N\mu_\infty^0 + 4\pi R^2\gamma$, where μ_∞^0 is the bulk energy per molecule and γ the interfacial free energy per unit area (Chapter 14). Hence

$$\mu_N^0 = \mu_\infty^0 + \frac{4\pi R^2\gamma}{N} = \mu_\infty^0 + \frac{4\pi\gamma(3v/4\pi)^{2/3}}{N^{1/3}} = \mu_\infty^0 + \frac{\alpha kT}{N^{1/3}}, \qquad (15.9)$$

where

$$\alpha = \frac{4\pi\gamma(3v/4\pi)^{2/3}}{kT} \approx \frac{4\pi r^2\gamma}{kT}, \qquad (15.10)$$

r being the effective radius of a molecule.

We see, therefore, that for the simplest shaped structures—rods, sheets, and spheres—the interaction free energy of the molecules can be expressed as

$$\mu_N^0 = \mu_\infty^0 + \alpha kT/N^p, \qquad (15.11)$$

where α is a positive constant dependent on the strength of the intermolecular interactions and p is a number that depends only on the shape or *dimensionality* of the aggregates. As we shall see, Eq. (15.11) also applies to various micellar structures and to spherical vesicles in which the bilayers bend elastically. In particular, we note that for all these structures, μ_N^0 decreases progressively with N, which is a necessary condition for aggregate formation.

15.5. The critical micelle concentration

Given the general functional form for μ_N^0 of Eq. (15.11) we may now ask, At what concentration will aggregates form? Incorporating Eq. (15.11) into the two fundamental equations of self-assembly, Eq. (15.2) and (15.4), leads us to some very interesting conclusions. First, we note that

$$X_N = N\{X_1 \exp[(\mu_1^0 - \mu_N^0)/kT]\}^N$$

$$= N\{X_1 \exp[\alpha(1 - 1/N^p)]\}^N \approx N[X_1 \exp \alpha]^N. \quad (15.12)$$

For sufficiently low monomer concentrations X_1 such that $X_1 \exp[(\mu_1^0 - \mu_N^0)/kT]$ or $X_1 \exp \alpha$ is much less than unity, we have $X_1 > X_2 > X_3 > \cdots$ for all α. Thus, at low concentrations most of the molecules in the solution will be as isolated monomers (i.e., $X_1 \approx C$) (Fig. 69). However, since X_N can never exceed unity it is clear from Eq. (15.12) that once X_1 approaches $\exp[-(\mu_1^0 - \mu_N^0)/kT]$ or $\exp(-\alpha)$ *it can increase no further*. The monomer

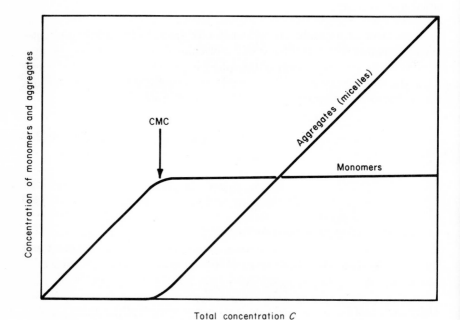

Total concentration C

Fig. 69 Monomer and aggregate concentrations as a function of total solute concentration (schematic). Most single-chained surfactants containing 12–16 carbons per chain have their CMC in the range 10^{-2}–10^{-5} mol dm^{-3}, while the corresponding double-chained surfactants have much lower CMC values due to their greater hydrophobicity (e.g., $\sim 5 \times 10^{-10}$ mol dm^{-3} for di-C_{16}-phosphatidylcholine) (Tanford, 1980).

concentration $(X_1)_{\text{crit}}$ at which this occurs may be called the *critical aggregation concentration* though we will use the more conventional term *critical micelle concentration* (CMC) to denote the critical aggregation concentration of all aggregated structures. Thus, in general

$$(X_1)_{\text{crit}} = \text{CMC} \approx \exp[-(\mu_1^0 - \mu_N^0)/kT], \tag{15.13}$$

while if μ_N^0 is given by Eq. (15.11),

$$(X_1)_{\text{crit}} = \text{CMC} \approx \exp(-\alpha) \qquad \text{for all } p. \tag{15.14}$$

These two equations define the concentration at which further addition of solute molecules results in the formation of more aggregates while leaving the monomer concentration more or less unchanged at the CMC value (Fig. 69).

15.6. Infinite aggregates (phase separation) versus finite aggregates (micellization)

What can we say about the nature of these aggregates? This now depends very much on their *shape*. For simple disc-like and spherical aggregates, Eq. (15.12) becomes

$$X_N = N[X_1 e^\alpha]^N e^{-\alpha N^{1/2}} \qquad \text{for discs } (p = \tfrac{1}{2}), \tag{15.15}$$

$$X_N = N[X_1 e^\alpha]^N e^{-\alpha N^{2/3}} \qquad \text{for spheres } (p = \tfrac{1}{3}), \tag{15.16}$$

which, above the CMC where $X_1 e^\alpha \approx 1$, may be approximated by $X_N \approx Ne^{-\alpha N^{1/2}}$ and $X_N \approx Ne^{-\alpha N^{2/3}}$, respectively. Now for any reasonable positive value of α, which is usually greater than 1, these equations show that apart from a few dimers, trimers, etc., *there will be very few aggregates of any appreciable size* (e.g., with $N > 5$). Where do the molecules go above the CMC? The answer is quite simple: For discs and spheres, there is a phase transition to a separate phase, strictly to an aggregate of infinite size $(N \to \infty)$, at the CMC. Israelachvili *et al.* (1976) showed that such a transition to large macroscopic aggregates occurs whenever $p < 1$ in Eq. (15.11). This applies, quite generally, to all planar or disc-like aggregates composed of identical molecules, and it is for this reason that finite crystalline sheets, one-component lipid bilayers, and even biological membranes with exposed edges are never found floating about in solution. Above the CMC infinite bilayers form spontaneously from lipid monomers, although they may close up on themselves to form vesicles (discussed later).

Likewise for simple spherical structures. Here, for example, we may consider the association of oil or alkane molecules dispersed in water where the molecules will remain as monomers up to a critical concentration given by

Eqs. (15.14) and (15.10),

$$(X_1)_{\mathrm{crit}} \approx e^{-\alpha} \approx e^{-4\pi r^2 \gamma / kT}, \tag{15.17}$$

above which they will separate out into a bulk oil phase, which may be considered simply as a very large spherical aggregate! $(X_1)_{\mathrm{crit}}$ for such systems (i.e., immiscible liquids) is conceptually the same as the CMC but is generally referred to as the *solubility* of the solute in the solvent, where α then represents the free energy of transferring a solute molecule from bulk solute into the solvent medium. For example, if we consider the solubility of hydrocarbons in water, we may put $\gamma \approx 50$ mJ m^{-2} and $r \approx 0.2$ nm for a methane molecule. Thus, the free energy of transferring a methane molecule from water into bulk hydrocarbon liquid should be approximately $4\pi r^2 \gamma \approx 2.5 \times 10^{-20}$ J ($\alpha \approx 6$) or about 15 kJ mol^{-1}. This theoretical estimate agrees well with the experimental value for the solubility or "hydrophobic energy" of methane. The hydrophobic free energy of transferring alkyl chains into bulk hydrocarbon or into micelles (which determines the CMC) can also be analysed in a similar fashion, though there is some uncertainty on where to define the hydrocarbon water "interface" and on the value of γ for such small molecules (Tanford, 1973, 1980; Herman, 1972, 1975). Experimentally one finds an increment of about 3.5 kJ mol^{-1} for each additional CH$_2$ group added to an alkyl chain. This corresponds to an increment in α of ~ 1.4 and to a lowering of the CMC by a factor of about $e^{-1.4} \approx 0.25$ per additional CH$_2$ group, though for some ionic micelles, the CMC is reduced by only 0.5 per added CH$_2$ group.

The important difference between alkanes and amphiphilic molecules is not so much in their solubility or CMC values but in the ability of amphiphiles to assemble into structures in which μ_N^0 reaches a minimum or constant value at some *finite* value of N, and for this reason the aggregates formed are not infinite ($\rightarrow$ phase separation) but of finite size ($\rightarrow$ micellization). The reason *why* amphiphilic molecules do this will be investigated fully in the following chapter.

Finally in this section we shall consider the properties of aggregates for which p $= 1$ in Eq. (15.11) (e.g., rod-like aggregates such as cylindrical micelles or linear chains of molecules) (Fig. 68). These have very different properties from the discs and spheres discussed above for reasons that will become immediately apparent. For rods, we must put $p = 1$ in Eq. (15.12) so that

$$X_N = N[X_1 e^\alpha]^N e^{-\alpha}. \tag{15.18}$$

In contrast to Eqs. (15.15) and (15.16) the second exponential term is now a constant rather than a rapidly decreasing function of N. Since above the CMC we have $X_1 e^\alpha \gtrsim 1$ we find that for small N, $X_N \propto N$ (i.e., there is no phase separation); the concentration of molecules in these aggregates now *grows* in proportion to their size! Only for very large N does the $[X_1 e^\alpha]^N$ term begin to dominate, eventually bringing X_N down to zero as N approaches infinity. Let us analyse this system in more detail.

The total concentration of molecules is given by inserting Eq. (15.18) into Eq. (15.4) as

$$C = \sum_{N=1}^{\infty} X_N = \sum_{N=1}^{\infty} N[X_1 e^\alpha]^N e^{-\alpha}$$

$$= e^{-\alpha}[X_1 e^\alpha + 2(X_1 e^\alpha)^2 + 3(X_1 e^\alpha)^3 + \cdots]$$

$$= X_1/(1 - X_1 e^\alpha)^2, \tag{15.19}$$

where we have made use of the identity

$$\sum_{N=1}^{\infty} Nx^N = x/(1 - x)^2.$$

Thus,

$$X_1 = \frac{(1 + 2Ce^\alpha) - \sqrt{1 + 4Ce^\alpha}}{2Ce^{2\alpha}}. \tag{15.20}$$

Note that at low concentrations C where $Ce^\alpha \ll 1$ this gives $X_1 \approx C$, whereas at high concentrations, well above the CMC such that $Ce^\alpha \gg 1$, the above simplifies to

$$X_1 \approx (1 - 1/\sqrt{Ce^\alpha})e^{-\alpha} \gtrsim e^{-\alpha} \tag{15.21}$$

(i.e., $X_1 \approx$ CMC as expected). Finally, above the CMC the density *distribution* of molecules in aggregates of N molecules is given by inserting the above equation back into Eq. (15.18):

$$X_N = N(1 - 1/\sqrt{Ce^\alpha})^N e^{-\alpha}. \tag{15.22}$$

This function peaks at $\partial X_N/\partial N = 0$, which occurs at

$$N \approx \sqrt{Ce^\alpha}, \tag{15.23}$$

while the *expectation value* of N, defined by $\langle N \rangle = \Sigma NX_N/\Sigma X_N = \Sigma NX_N/C$, is given by

$$\langle N \rangle = \sqrt{1 + 4Ce^\alpha}$$

$$\approx 1 \qquad \text{below the CMC,} \tag{15.24}$$

$$\approx 2\sqrt{Ce^\alpha} \qquad \text{above the CMC.}$$

Thus, the distribution is very broad, or *polydisperse*, with X_N increasing linearly with N for small aggregates and decaying gradually to zero only at large N ($N \gg \sqrt{Ce^\alpha}$). Note that for small N, the concentration of *aggregates*, X_N/N, remains roughly constant.

The above results should apply to all dilute aggregates for which p = 1 in Eq. (15.11) (e.g., cylindrical micelles, fibrous structures such as microfilaments and microtubules, as well as linear arrays of particles often seen in membranes). Later we shall see that it also applies to spherical vesicles whose membranes bend elastically. For all these structures, the mean aggregation number is concentration-dependent, varying with the square root of the concentration C above the CMC, and from Eq. (15.24) we note that it is also very sensitive to small changes in the interaction parameter α. Consequently, we may anticipate that the aggregation number and polydispersity of rod-like structures in water should be very sensitive to temperature, electrolyte concentration, and pH.

15.7. More complex amphiphilic structures

So far the thermodynamic analysis has been concerned with very simple aggregates—rods, sheets, and spheres—without considering *why* different molecules should aggregate into one or another of these structures. This is determined by the type of intermolecular binding forces operating between the molecules. For symmetric molecules such as alkanes in water whose hydrophobic interaction with each other is *nondirectional*, we would expect them to coalesce and grow as small spherical droplets (for which the total surface energy is a minimum for any given N). And we have seen that for such aggregates, $p = \frac{1}{3}$ in Eq. (15.11), which results in a phase separation at the solubility limit (the effective CMC). Clearly, molecules that aggregate into linear or sheet-like structures must have asymmetric *directional* bonding (e.g., oriented at either end of or axially around the molecules for them to link up into one and two dimensional aggregates, respectively). Further, if the molecules are also flexible, the structures they adopt may be more varied than the simple shapes so far considered. Thus, the energetically unfavourable regions at each end of a rod-like aggregate may be eliminated if the two ends bend and join together, resulting in a torus. Likewise, the unfavourable rim energy of a disc may be eliminated by its closing up into a vesicle, which is what happens in surfactant and lipid assemblies. In such cases μ_N^0 reaches a minimum value at some *finite* value of N (say $N = M$) or reaches a low value at $N = M$ and then remains almost constant for $N > M$. As we shall see this also occurs for spherical micelles.

Depending on the sharpness of the minimum in the free energy, such a form for μ_N^0 results in monodisperse aggregates of mean aggregation number $N \gtrsim M$ (rather than infinite or polydisperse aggregates), with the CMC occurring at

$$\text{CMC} \approx \exp[-(\mu_1^0 - \mu_M^0)/kT]. \tag{15.25}$$

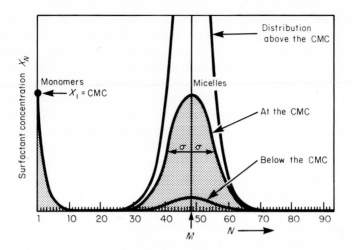

Fig. 70 Distribution of molecules X_N as a function of aggregation number N. Near the CMC (shaded region) we have $X_1 \approx X_M$ where the mean micellar aggregation number is M. For spherical micelles, the distribution about M is near Gaussian with standard deviation $\sigma \approx \sqrt{M}$ (see Section 16.4).

This is the common form traditionally used in the analysis of spherical micelles. If μ_N^0 has a minimum value at $N = M$, the variation of μ_N^0 about μ_M^0 can usually be expressed in the parabolic form:

$$\mu_N^0 - \mu_M^0 = \Lambda(\Delta N)^2, \tag{15.26}$$

where $\Delta N = (N - M)$. In this case Eq. (15.3) becomes

$$X_N = N\left\{\frac{X_M}{M}\exp[-M\Lambda(\Delta N)^2/kT]\right\}^{N/M}, \tag{15.27}$$

and so the distribution in X_N about M will be near Gaussian (Fig. 70) with a standard deviation given by

$$\sigma = \sqrt{kT/2M\Lambda}. \tag{15.28}$$

Systems that fall into this category are spherical micelles and single bilayer vesicles.

15.8. Effects of interactions between aggregates: mesophases and multilayers

So far we have ignored interaggregate interactions. These cannot be ignored at high concentrations (low water content) where, especially for surfactant and lipid dispersions, transitions to larger and more ordered *mesophase* structures are commonly observed. Both attractive and repulsive forces between aggregates can lead to such *phase transitions*. For example, if there are strong long-range *repulsive* electrostatic or hydration forces between aggregates, then above some surfactant concentration spherical micelles will try to get as far apart from each other as possible, but being constrained in a finite solution they cannot do this. However, if the surfactants rearrange to form an ordered array of cylindrical structures, their surfaces can now be farther apart from each other. And if they order into a stack of bilayers, their surfaces can be even farther apart, all at the same surfactant concentration. It is for this reason that many surfactant structures go from being spherical micelles to infinite cylinders to multibilayers (lamellae) as the surfactant content is progressively increased above about 10% by weight (Ekwall, 1975; Tiddy, 1980).

Here we shall consider the transitions of suspensions of micelles and/or vesicles into multilayer structures (liposomes), brought about when the *attractive* van der Waals forces between them dominate their long-range interaction (i.e., in the absence of long-range repulsive forces). This can occur for nonionic or zwitterionic head groups as well as for charged head groups in high salt where the electrostatic repulsion is screened. In such cases the equilibrium separation between the bilayers in the liposomes will be given by the position of the potential-energy minimum, the depth of the minimum being given by W per unit area (cf. Figs. 47 and 78). Now while smaller aggregates such as vesicles are clearly favoured entropically, larger aggregates such as liposomes could be thermodynamically more favourable if the value of W is sufficiently large. The problem is to establish how these two effects compete in determining which structure is formed at the CMC and at higher concentrations.

If M is the micelle or vesicle aggregation number and $\mathbf{M}$ the multilayer aggregation number $(\mathbf{M} \gg M)$, then equating the chemical potentials of molecules in all the possible dispersed and aggregated states gives at equilibrium

$$\mu_1^0 + kT \log X_1 = \mu_M^0 + (kT/M)\log(X_M/M) = \mu_\mathbf{M}^0 + (kT/\mathbf{M})\log(X_\mathbf{M}/\mathbf{M})$$
$$\text{monomers} \qquad \text{micelles/vesicles} \qquad \text{liposomes}$$

$$(15.29)$$

or

$$(X_\mathbf{M}/\mathbf{M}) = \{(X_M/M)\exp[M(\mu_M^0 - \mu_\mathbf{M}^0)/kT]\}^{\mathbf{M}/M}. \qquad (15.30)$$

The concentration at which $X_\mathbf{M} = X_M$ is therefore

$$(X_M)_{\text{crit}} \approx M\exp[-M(\mu_M^0 - \mu_\mathbf{M}^0)/kT]. \qquad (15.31)$$

Thus, depending on M and the difference in the free energies $(\mu_M^0 - \mu_\mathbf{M}^0)$ between molecules in the micellar and liposome states (which includes contributions from both *intra*aggregate and *inter*bilayer interactions) the latter may form spontaneously at the CMC, or—if $(X_M)_{\text{crit}}$ exceeds the CMC—at some higher concentration, while the background concentration of monomers and the smaller aggregates remains unchanged. We may conveniently term such transitions first and second CMCs. If $\mathbf{M} \gg M$, the concentration at which large aggregates begin to form will be sharp and in all respects analogous to a first CMC. Note, too, that if we put $M = 1$ in Eq. (15.31), it reduces to Eq. (15.13) for the first CMC. Indeed, if we consider the smaller aggregates as if they were "monomers," Eqs. (15.30) and (15.31) are completely analogous to Eqs. (15.2) and (15.13).

As an example, let us consider a suspension of vesicles in equilibrium with a lamellar (multilayer) phase, where the depth of the potential energy minimum between bilayers in the lamellar phase is $W = 8 \times 10^{-2}$ mJ m^{-2} and the surface area occupied by each lipid molecule is 0.70 nm^2. Thus, the free energy difference *per molecule* in vesicles relative to that in the multilayers is

$$(\mu_M^0 - \mu_{\mathbf{M}}^0) \approx 4 \times 10^{-5} \times 0.70 \times 10^{-18} = 2.8 \times 10^{-23} \quad \text{J,}$$

or $0.0068\ kT$ at 298 K. If $M = 3000$ is the vesicle aggregation number, then from Eq. (15.31) a vesicle-to-multilayer transition will occur at a critical lipid concentration of

$$(X_M)_{\text{crit}} = 3000 \exp[-3000 \times 0.0068] \approx 4 \times 10^{-6},$$

that is, at about 2×10^{-3} mol dm^{-4} (which may be compared with the first CMC of about 10^{-10} mol dm^{-3} typical of many vesicle-forming lipids). This analysis, however, neglects any possible energy difference arising from the different bilayer curvatures in vesicles and planar bilayers (see Section 16.9), which will also contribute to $(\mu_M^0 - \mu_{\mathbf{M}}^0)$ in Eq. (15.31).

15.9. Conclusion

We have gone as far as is possible with a formal analysis of the statistical thermodynamics of self-assembly, and to proceed further we must now consider the different types of interactions occurring between amphiphilic molecules in aggregates more specifically. In Chapter 16 we shall quantify these interactions and in particular investigate how the geometric shape constraints of surfactant and lipid molecules restrict their assembly into aggregates of different shapes. The important conclusion to be drawn from this chapter is that once the aggregates' shape is known their physical properties are necessarily given by the thermodynamic equations developed in this chapter, for example, a dilute dispersion of (one-component) cylindrical micelles *must* be polydisperse and their size *must* increase with concentration. Therein lies the beauty and power of thermodynamics.

Chapter 16

Aggregation of Amphiphilic Molecules into Micelles, Bilayers, Vesicles, and Biological Membranes

16.1. Introduction

Amphiphilic molecules such as surfactants and lipids can associate into a variety of structures in aqueous solutions (Fig. 65), which can transform from one to another when the solution conditions are changed, for example, the electrolyte concentration or pH. In order to understand these structural aspects one requires an understanding not only of the thermodynamics of self-association (discussed in Chapter 15) but also of how the interaction forces between amphiphiles *within* aggregates are affected by solution conditions. These two factors, and the strength of the forces *between* aggregates in more concentrated systems, determine the equilibrium structures formed. In this chapter we shall investigate the interactions between amphiphilic molecules within aggregates in more detail, and we shall see how the concept of molecular geometry or molecular packing constraints allows us to understand the structures formed.

16.2. Optimal head-group area

The major forces that govern the self-assembly of amphiphiles into well-defined structures such as micelles and bilayers derive from the hydrophobic interaction at the hydrocarbon–water interface, which induces the molecules

246

to associate, and the hydrophilic nature of the head groups, which imposes the opposite requirement that they remain in contact with water. These two interactions compete to give rise to the idea of two "opposing forces" (Tanford, 1980) acting mainly in the interfacial region: the one tending to decrease and the other tending to increase the interfacial area a per molecule (the head-group area) exposed to the aqueous phase (Fig. 71). The attractive interaction arises mainly from the attractive hydrophobic or interfacial tension forces acting at the essentially *fluid* hydrocarbon–water interface and may be represented by a positive interfacial free energy per unit area characteristic of such interfaces of $\gamma \approx 50$ mJ m^{-2}, though there are indications that this value may be much smaller and closer to 20 mJ m^{-2} (Parsegian, 1966; Jönsson and Wennerström, 1981). Thus the interfacial hydrophobic free energy contribution to μ_N^0 may be written as γa.

The repulsive contributions are too complex and difficult to formulate explicitly; these include the electrostatic (double-layer) repulsion between charged head groups (Payens, 1955; Forsyth *et al.*, 1977), hydration repulsion,

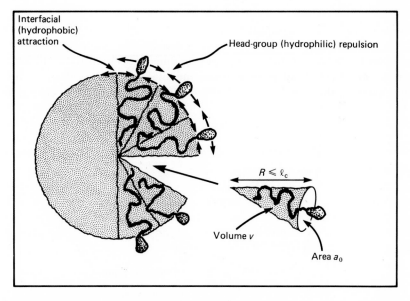

Fig. 71 The hydrocarbon interiors in both micelles and bilayers are normally in the fluid state (Shinitzky *et al.*, 1971; Lindblom and Wennerström, 1977). Repulsive head-group forces and attractive hydrophobic interfacial forces determine the optimum head-group area a_0 at which μ_N^0 is a minimum (see Fig. 72). The chain volume v and chain length l_c set limits on how the fluid chains can pack together, on average, inside an aggregate (i.e., the mean molecular conformation depends on a_0, v, and l_c).

and steric head-group and chain interactions (Israelachvili *et al.*, 1980a). However, as in the two-dimensional van der Waals equation of state, we expect the first term in any energy expansion to be inversely proportional to the surface area occupied per head group a [cf. pressure $\propto 1/a^2$ in Eq. (6.15)].

The total interfacial free energy per molecule in an aggregate may therefore be written, to first order, as

$$\mu_N^0 = \gamma a + K/a, \qquad (16.1)$$

where K is a constant. We shall initially assume that both these forces act in the same plane. The minimum free energy is therefore given when $\partial \mu_N^0 / \partial a = 0$, that is,

$$\mu_N^0(\text{min}) = 2\gamma a_0, \qquad a_0 = \sqrt{K/\gamma}. \qquad (16.2)$$

a_0 will be referred to as the *optimal surface area* per molecule defined at the hydrocarbon–water interface. The interfacial energy per molecule, Eq. (16.1), may now be expressed in the more convenient form

$$\mu_N^0 = 2\gamma a_0 + \frac{\gamma}{a}(a - a_0)^2 \qquad (16.3)$$

in which the unknown constant K has been eliminated, so that μ_N^0 as a function of a is now in terms of the two known or measurable parameters γ and a_0.

We see therefore how the concept of opposing forces leads to the notion of an optimal area per head group at which the total interaction energy per lipid molecule is a minimum (Fig. 72). Moreover, for fluid hydrocarbon chains, the

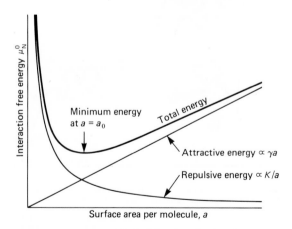

Fig. 72 Optimal head-group area a_0 at which the opposing forces of head-group repulsion and interfacial (hydrophobic) attraction are balanced.

optimal area should not depend on the chain length, as is indeed found experimentally (Gallot and Skoulios, 1966; Reiss-Husson, 1967; Lewis and Engelman, 1983b).

The above equations, while crude, contain the essential features of interlipid interactions in micelles, bilayers, and membranes. They imply that the interaction energy between lipids has a minimum at a certain head-group area a_0, about which the energy varies parabolically (i.e., elastically). The equations ignore specific head-group interactions, ion binding effects, as well as chain–chain interactions within the hydrocarbon core (i.e., the hydrocarbon chains are assumed to be perfectly fluid).

16.3. Geometric packing considerations

Having established the equations that adequately describe the interaction between lipids within aggregates, we have yet to establish the type of structure into which different lipids will assemble. Geometric considerations must now be applied to determine the most favoured structure (i.e., the shapes and sizes of molecules now enter the picture). The geometric or packing properties of lipids depend on their optimal area a_0, the volume v of their hydrocarbon chain or chains, which will be assumed to be fluid and incompressible, and the maximum effective length that the chains can assume. We shall call this the *critical chain length* l_c. This length sets a limit on how far the chains can extend; smaller extensions are allowed but further extensions are not, these being prevented by a sharp rise in free energy. The critical length l_c is a semi-empirical parameter, since it represents a somewhat vague cutoff distance beyond which hydrocarbon chains can no longer be considered as fluid (see also Section 16.9). However, as may be expected it is of the same order as, though somewhat less than, the fully extended molecular length of the chains l_{max} (Tanford, 1973, 1980; Israelachvili *et al.*, 1976, 1977). According to Tanford,

$$v \approx (27.4 + 26.9n) \times 10^{-3} \quad \text{nm}^3 \tag{16.4}$$

and

$$l_c \gtrsim l_{max} \approx (0.15 + 0.1265n) \quad \text{nm} \tag{16.5}$$

per saturated hydrocarbon chain with n carbon atoms. Note that for large n, $v/l_c \approx 0.21 \text{ nm}^2 \approx$ constant, which is the minimum cross-sectional area that a hydrocarbon chain can have.

Once the optimal surface area a_0, hydrocarbon chain volume v, and critical length l_c are specified for a given lipid—all these being measurable or estimable—it is possible to determine the shape and size of the structure into which the lipids can pack while maintaining these parameters. It turns out that these can be satisfied by a great variety of different structures. However, since μ_N^0 will be roughly the same for all these structures (since a_0 is the same) entropy will favour the structure with the smallest aggregation number, say, at $N = M$, and this structure is unique! Larger structures will be entropically unfavoured while smaller structures, where packing constraints cause the surface area a to increase above a_0, will be energetically unfavoured.

It has been shown previously (Israelachvili et al., 1976) that for lipids of optimal area a_0, hydrocarbon volume v, and critical chain length l_c, the value of the dimensionless packing parameter $v/a_0 l_c$ will determine whether they will form spherical micelles ($v/a_0 l_c < \frac{1}{3}$), nonspherical micelles ($\frac{1}{3} < v/a_0 l_c < \frac{1}{2}$), vesicles or bilayers ($\frac{1}{2} < v/a_0 l_c < 1$), or inverted micelles ($v/a_0 l_c > 1$). Each of these structures corresponds to the minimum-sized aggregate in which all the lipids have minimum free energy (i.e., $a = a_0$). These structures will now be described in turn.

16.4. Spherical micelles

For lipids to assemble into spherical micelles, their optimal surface area a_0 must be sufficiently large and their hydrocarbon volume v sufficiently small such that the radius of the micelle will not exceed the critical chain length l_c. From simple geometry we have, for a spherical micelle of radius R and mean aggregation number M (Fig. 71),

$$M = 4\pi R^2/a_0 = 4\pi R^3/3v, \tag{16.6}$$

namely,

$$R = 3v/a_0, \tag{16.7}$$

so that only if $R \leq l_c$, that is, only for

$$v/a_0 l_c \leq \frac{1}{3}, \tag{16.8}$$

will the amphiphiles be able to pack into a spherical micelle with their head-group areas equal to a_0 and with the micelle radius not exceeding l_c.

An example of such micelle-forming amphiphiles is the 12-carbon chain sodium dodecyl sulphate surfactant (SDS) in water (Fig. 73) where experimentally $M \approx 57$, $v = 0.350$ nm^3 [Eq. (16.4)], and therefore from Eq. (16.6) $a_0 \approx 0.62$ nm^2 and $R \approx 1.7$ nm. Now for a 12-carbon chain, Eq. (16.5) gives $l_c \approx 1.7$ nm; thus $v/a_0 l_c \approx \frac{1}{3}$ for SDS micelles in water.

The mean size of spherical micelles is relatively insensitive to the surfactant concentration above the CMC, and the micelles are fairly monodisperse. The standard deviation σ in the aggregation number about the mean (at $N \approx M$)

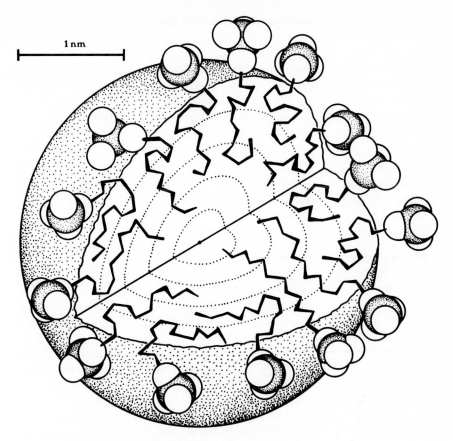

1 nm

Fig. 73 A sodium dodecylsulphate (SDS) micelle drawn to scale. The micelle contains 60 sodium dodecylsulphate molecules. The hydrocarbon chains pack at liquid hydrocarbon density in the core where they are almost as disordered as in the bulk liquid state. Each of the five spherical shells contains approximately the correct number of chain segments to ensure even chain packing density throughout. Note that all segments of the chain spend an appreciable proportion of time near the micelle surface. Thus, even though the core is almost completely devoid of water each segment samples the hydrophilic environment. Drawing based on calculations by Gruen (1981) and Gruen and de Lacey (1984).

may be obtained by first noting that μ_N^0 of Eq. (16.3) may be written as

$$\mu_N^0 = \mu_M^0 + \frac{\gamma}{a}(a - a_0)^2. \tag{16.9}$$

Now for a spherical micelle, we have

$$N = 4\pi R^2/a = 4\pi R^3/3v = 36\pi v^2/a^3,$$

where $N = M$ for $a = a_0$. Equation (16.9) may therefore be written as

$$\mu_N^0 - \mu_M^0 = \Lambda(N - M)^2, \tag{16.10}$$

where $\Lambda = \gamma a_0/9M^2$. Substituting this value for Λ into Eq. (15.28) gives

$$\sigma = \sqrt{(9kT/2\gamma a_0)M}. \tag{16.11}$$

Typically, for γ lying in the range 20–50 mJ m^{-2} and $a_0 \approx 0.60$ nm^2, we expect

$$\sigma \approx \sqrt{M}, \tag{16.12}$$

that is, for $M \approx 60$, $\sigma \approx 8$. Aniansson *et al.* (1976) found that for a variety of sodium alkyl sulfate micelles, the value of $\sigma/\sqrt{M}$ varies between 1 and 2. The distribution or spread about M is thus fairly narrow but by no means sharp, as illustrated in Fig. 70.

16.5. Nonspherical and cylindrical micelles

Most lipids that form spherical micelles have charged head groups since this leads to large head-group areas a_0. Addition of salt partially screens the electrostatic head-group repulsion and thereby reduces a_0. Those lipids that possess smaller head-group areas such that $\frac{1}{3} < v/a_0 l_c < \frac{1}{2}$ cannot pack into spherical micelles but can form cylindrical (rod-like) structures. Thus, single-chained lipids possessing charged head groups in high salt or those possessing uncharged (nonionic or zwitterionic) head groups fall into this category (e.g., SDS and CTAB in high salt, lysolecithins).

As discussed in Section 15.6 rod-like aggregates must have very unusual properties: They are large and polydisperse, and their mean aggregation number is very sensitive to the total lipid concentration C. According to Eq. (15.24) their mean aggregation number $\langle N \rangle$ should increase proportionally to $\sqrt{C}$ above the CMC, which is indeed found to be the case experimentally (Mazer *et al.*, 1976). It is important to note that the unusual properties of cylindrical micelles are due entirely to end effects: At each end the lipids *have to* pack into hemispherical caps with head-group area a given by $v/al_c = \frac{1}{3}$, so that $a > a_0$ since $v/a_0 l_c > \frac{1}{3}$. The unfavourable energy of these end lipids determines the magnitude of the interaction parameter α in Eq. (15.24), which increases as a_0 decreases. Since $\langle N \rangle = 2\sqrt{C e^\alpha}$ it is not surprising that the growth of cylindrical micelles is very sensitive to changes in temperature, chain length, and—for ionic lipids–to ionic strength (Missel *et al.*, 1983). For example, their sensitivity to increasing ionic strength arises from its effect on decreasing a_0, which increases the interaction parameter α. As might be expected the mean aggregation number increases dramatically with increased ionic strength. Thus, the aggregation number of SDS micelles in

0.6 M NaCl is ~ 1000 compared to ~ 60 in water (Wennerström and Lindman, 1979). The unfavourable end energy of cylindrical micelles may be eliminated if the two ends join, thereby forming a toroidal micelle (the two-dimensional analogue of a vesicle). However, toroidal micelles have so far not been observed.

16.6. Bilayers

Lipids that form bilayers are those that cannot pack into small micellar structures due to their small head-group area a_0 or, as is more common, because their hydrocarbon chains are too bulky to fit into such small aggregates while maintaining the surface area at its optimal value. For bilayer-forming lipids, the value of $v/a_0 l_c$ must lie close to 1, and this requires that for the same head-group area a_0 and chain length l_c, their hydrocarbon volume v must be about twice that of micelle-forming lipids (for which $v/a_0 l_c$ is normally in the range $\frac{1}{3}$ to $\frac{1}{2}$). Therefore lipids with *two* chains are likely to form bilayers, and indeed most of them do. For example, lysolecithin forms small but nonspherical micelles while lecithins with two alkyl chains (e.g., di-C_{14} and di-C_{16} lecithins) form bilayers (Fig. 74). However, the doubling of the chains also increases the hydrophobicity of the lipids, which drastically lowers their CMC (compare the CMCs of common micelle-forming lipids with 12–16 carbons, 10^{-2}–10^{-5} M, with those of bilayer-forming lipids, $\sim 10^{-10}$ M). We shall return to comment on the biological significance of such low CMCs later.

When a lipid bilayer is stretched from its equilibrium state it expands elastically. The area compressibility modulus k_c may be readily estimated from Eq. (16.3) since by definition

$$\text{elastic energy} = \tfrac{1}{2}k_c(a - a_0)^2/a, \tag{16.13}$$

which gives a value of $k_c \approx 2\gamma \approx 100$ mJ m^{-2} per monolayer (assuming $\gamma \approx 50$ mJ m^{-2}), while for bilayers, $k_c \approx 4\gamma \approx 200$ mJ m^{-2}. This estimate falls in the range of measured k_c values on fluid lipid bilayers and free biological cell membranes (100–230 mJ m^{-2}).

16.7. Vesicles

Under certain conditions it becomes more favourable for closed spherical bilayers (vesicles) to form rather than infinite planar bilayers (Fig. 65). This arises since in a closed bilayer the energetically unfavourable edges are eliminated at a finite, rather than infinite, aggregation number, which is entropically favoured. Thus, so long as the curved bilayer lipids can maintain their areas at their optimal value vesicles should be the preferred structures.

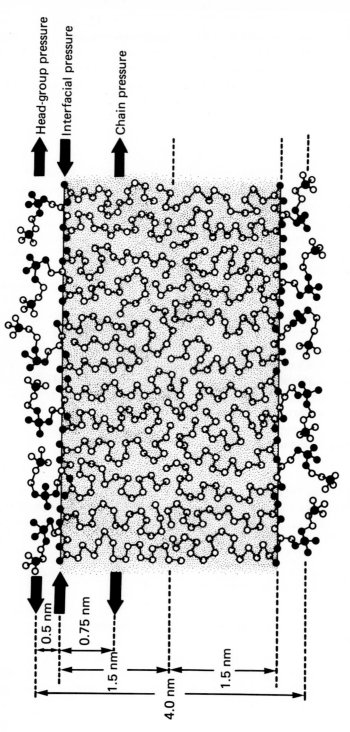

Fig. 74 Lecithin bilayer drawn to scale. The lipid bilayer is the basic structure of biological membranes, and most membrane lipids contain two hydrocarbon chains. The lipids diffuse rapidly in the plane of the bilayer, covering a distance of about 1 μm in 1 s. They also cross the bilayer from one side to the other ("flip-flop"), as well as exchange with lipids in the solution, but at much slower rates, of the order of hours (cf. the lifetime of single-chained surfactants in micelles of 10^{-5} to 10^{-3} s). (From Israelachvili *et al.*, 1980a.)

What then determines the radii of vesicles? First, let us note that if $v/a_0 l_c = 1$, only planar bilayers will form. For a bilayer to curve, the lipids in the outer monolayer must be able to pack, on average, into truncated cones. This requires that $v/a_0 l_c < 1$. Simple geometric considerations show (Israelachvili et al., 1976) that for $\frac{1}{2} < v/a_0 l_c < 1$, the radius of the smallest vesicle that may be formed without forcing the head-group area a in the outer monolayer to exceed a_0 is given by

$$R_c \approx l_c \left[\frac{3 + \sqrt{3(4v/a_0 l_c - 1)}}{6[1 - v/a_0 l_c]} \right] \approx \frac{l_c}{(1 - v/a_0 l_c)}, \qquad (16.14)$$

which is the *critical radius* below which a bilayer may not curve without introducing unfavourable packing strains on the lipids. So long as a vesicle's radius does not fall below R_c the lipids in both the inner and outer monolayers can pack with their surface areas at the optimal value a_0 and with the two hydrocarbon chain regions not exceeding l_c. For $R < R_c$, a must exceed a_0 in the *outer* monolayer, and such vesicles are energetically unfavoured. For a vesicle of radius R_c and bilayer hydrocarbon thickness $t \approx 2v/a_0$, the aggregation number is

$$N \approx 4\pi[R_c^2 + (R_c - t)^2]/a_0. \qquad (16.15)$$

As an example we may apply the above equations to the much-studied egg lecithin vesicles for which $a_0 \approx 0.717$ nm^2 and $v \approx 1.063$ nm^3. Taking $l_c \approx 1.75$ nm (i.e., $v/a_0 l_c \approx 0.85$), Eqs. (16.14) and (16.15) then yield $R_c \approx 11$ nm, $N \approx 3000$, $t \approx 3.0$ nm, and an outside-to-inside lipid ratio of $R_c^2/(R_c - t)^2 \approx 1.9$, all in good agreement with measured values.

Many people now believe that vesicles represent the most likely prototypes of early living cells, and it has now been demonstrated that biological lipids as well as synthetic surfactants containing two hydrocarbon chains can spontaneously self-assemble into stable monodisperse vesicles (Brunner *et al.*, 1976; Hauser and Gains, 1982; Hauser, 1984; Talmon *et al.*, 1983).

Finally, we may note that for lipids with very small optimal head-group areas (e.g., $a_0 < 0.42$ nm^2 for double-chained lipids) or for bulky polyunsaturated chains, their value of $v/a_0 l_c$ will exceed unity. Such lipids cannot even pack into planar bilayers since their head-group area is too small; instead, they form inverted micellar structures or precipitate out of solution (e.g., MGDG, *cis*-unsaturated phosphatidylethanolamine, cholesterol).

16.8. Factors effecting changes from one structure to another

We have seen that the geometric packing properties of different lipids may be conveniently expressed in terms of the dimensionless *critical packing*

TABLE XXI

Mean (dynamic) packing shapes of lipids and the structures they form

Lipid	Critical packing parameter $v/a_0 l_c$	Critical packing shape[b]	Structures formed
Single-chained lipids (surfactants) with large head-group areas: *SDS in low salt*	$< 1/3$	Cone	Spherical micelles
Single-chained lipids with small head-group areas: *SDS and CTAB in high salt, nonionic lipids*	1/3-1/2	Truncated cone	Cylindrical micelles
Double-chained lipids with large head-group areas, fluid chains: *Phosphatidyl choline (lecithin), phosphatidyl serine, phosphatidyl glycerol, phosphatidyl inositol, phosphatidic acid, sphingomyelin, DGDG[a], dihexadecyl phosphate, dialkyl dimethyl ammonium salts*	1/2-1	Truncated cone	Flexible bilayers, vesicles
Double-chained lipids with small head-group areas, anionic lipids in high salt, saturated frozen chains: *phosphatidyl ethanolamine, phosphatidyl serine + Ca^{2+}*	~1	Cylinder	Planar bilayers
Double-chained lipids with small head-group areas, nonionic lipids, poly *(cis)* unsaturated chains, high T: *unsat. phosphatidyl ethanolamine, cardiolipin + Ca^{2+} phosphatidic acid + Ca^{2+} cholesterol, MGDG[b]*	> 1	Inverted truncated cone or wedge	Inverted micelles

[a] DGDG, digalactosyl diglyceride, diglucosyl diglyceride; [b] MGDG, monogalactosyl diglyceride, monoglucosyl diglyceride.

parameter $v/a_0 l_c$ characteristic for each lipid, the value of which determines the type of aggregate formed. Table XXI illustrates the structures formed by some common lipids, and how these can be modified by their ionic environment, temperature, unsaturation, etc., as will now be summarized for lipids with two hydrocarbon chains.

(i) Lipids with smaller head-group areas (high $v/a_0 l_c$) should form larger vesicles, less-curved bilayers, or inverted micellar phases. For anionic lipids, this can be brought about by increasing the salt concentration, particularly Ca^{2+}, or lowering the pH. This also has the effect of straightening (condensing) the chains.

(ii) Increased unsaturation, particularly of *cis* double bonds, reduces l_c and thus increases $v/a_0 l_c$. This leads to larger vesicles and ultimately to inverted structures.

(iii) For double-chained lipids above their chain melting temperature T_c, increasing the temperature increases the hydrocarbon chain motion, involving *trans–gauche* isomerization, and thereby reduces their limiting length l_c. If the head-group area a_0 does not increase, this again leads to an increased $v/a_0 l_c$.

The above effects of head-group size, ionic strength, unsaturation, and temperature are illustrated in Table XXI for dilute lipid structures in aqueous solutions. It is worth noting that micelles and bilayers also form in other hydrogen-bonding liquids, such as hydrazine (N_2H_4) and ethylene glycol, but their properties have not been as extensively studied.

16.9. Curvature elasticity of bilayers and membranes

The above treatment offers a rough and ready recipe for analysing the packing properties of lipid structures, but it is nevertheless incomplete. The only restriction we have so far considered is that the chains cannot extend away from the head group further than a certain distance l_c. This leads to an increased energy only for radii of curvature below R_c, but for $R > R_c$, there is no curvature dependence at all because of two assumptions that now require further consideration.

(i) It was assumed that hydrocarbon chains are entirely fluid and do not oppose any distortion (until they become forced to extend beyond l_c). In other words, it was assumed that there is no curvature dependence of μ_N^0 due to the nonfluidity of the chains. Gruen and de Lacey (1984), inspired by the earlier theory of Marčelja (1974), analysed the energetics of chain packing in both micelles and bilayers and found that the assumption of fluidity in bilayers is much less valid than it is in micelles. They concluded that $l_c \approx l_{max}$ in spherical and cylindrical micelles, but that $l_c \approx 0.7 l_{max}$ in bilayers (cf. the assumed value of $l_c \approx 1.75$ nm $\approx 0.75 l_{max}$ for egg lecithin vesicles in Section 16.7). Further, the restriction of chain freedom in bilayers gives rise to a lateral chain pressure (Fig. 74).

(ii)   The attractive and repulsive interfacial forces that determine a_0 and μ_N^0 (Section 16.2) were both assumed to act in the same plane, at the hydrocarbon–water interface, at which the surface area per molecule has been defined. This is likely to be true for the attractive interfacial tension force but not for the head-group repulsive forces, which are likely to be centred above the interface (Figs. 71 and 74). Again, this effect leads to a curvature dependence of μ_N^0.

The first effect acts to oppose bending right from the start and thus gives rise to an effective *positive* bilayer elastic curvature modulus, while the second effect favours bending and gives rise to a *negative* curvature modulus (Israelachvili *et al.*, 1976, 1980a). Let us consider these two cases in turn.

Positive curvature modulus. If k_b is the positive curvature or bending modulus of a membrane its elastic energy per unit area is defined by $\frac{1}{2}k_b/R^2$, where R is the radius of curvature. Thus, the interaction energy per vesicle of large radius R and aggregation number N is simply

$$N\mu_N^0 = N\mu_\infty^0 + (\tfrac{1}{2}k_b/R^2)4\pi R^2 = N\mu_\infty^0 + 2\pi k_b, \qquad (16.16)$$

and thus

$$\mu_N^0 = \mu_\infty^0 + 2\pi k_b/N, \qquad (16.17)$$

which is the same as Eq. (15.6) with $\alpha = 2\pi k_b/kT$. Thus, the mean aggregation number $\langle N \rangle$ of elastic bilayer vesicles will be given by Eq. (15.24) as

$$\langle N \rangle = 2\sqrt{Ce^\alpha} = 2e^{\pi k_b/kT}\sqrt{C}, \qquad (16.18)$$

where C is the lipid concentration (in mol dm^{-3}/55.5). Thus, for typical dilute lipid concentrations of 10^{-3} mol dm^{-3} and below, we see that if $k_b < 1.5 \times 10^{-20}$ J ($\alpha < 25$), vesicle populations will be small and monodisperse, with mean radii $R \approx R_c$. But if $k_b > 1.5 \times 10^{-20}$ J, vesicles will be large and polydisperse, and their size will increase with the total lipid concentration.

Eventually, at high enough concentrations large vesicles must always transform into extended bilayers, either because of space requirements or because of the attractive forces between them, as discussed in Section 15.8.

Negative curvature modulus. A negative elasticity favours bending of a bilayer right from the start and leads to smaller vesicles than expected from simple packing considerations. The effect is enhanced for large repulsive head groups, as well as for shorter hydrocarbon chains. Thus, below a certain chain length no stable vesicles should form; instead cylindrical or spherical micelles become the preferred structures as occurs for double-chained lecithins with 12 or fewer carbons per chain.

16.10. Biological membranes

Membranes are the most common cellular structures in both animals and plants (Fig. 75) where they are involved in almost all aspects of cellular activity ranging from simple mechanical functions such as motility, food entrapment, and transport to highly specific biochemical processes such as energy transduction, immunological recognition, nerve conduction, and biosynthesis (for

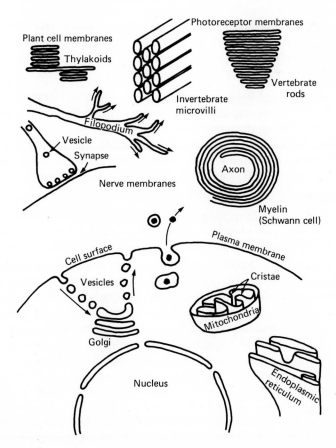

Fig. 75 Cellular membranes are thin sheets of lipids and proteins. Most biological membranes offer little resistance to bending.

text books on membranes in general see Robertson, 1983; Harrison and Lunt, 1980; Finean *et al.*, 1978; Quinn, 1976). Biological membranes are very complex and varied. They commonly contain 50 or more different proteins and a host of phospholipids and glycolipids with various head groups, numbers of chains, chain lengths, and degrees of unsaturation, as well as steroids (e.g., cholesterol) and other amphiphilic molecules. Yet in spite of their complexity there are many aspects of membrane structure that may be qualitatively understood in terms of the concepts we have already outlined, and it is best to start with a consideration of membrane lipids.

16.11. Membrane lipids

Most biological membrane lipids are *double-chained* phospholipids or glycolipids, with *16 to 18* carbons per chain, one of which is *unsaturated* or *branched* (see Table XX). These properties are not accidental but carefully designed by nature to ensure

(i) that biological lipids will self-assemble into thin bilayer membranes that can compartmentalize different regions within a cell as well as protect the inside of the cell from the outside;

(ii) that because of their extremely low CMC the membranes remain intact even when the bathing medium is grossly depleted of lipids; and

(iii) that because of the unsaturation or branching the membranes are in the fluid state at physiological temperatures.

For example, the chain melting temperatures T_c of saturated di-C_{16} phosphatidyl choline and phosphatidyl serine are 41 and 53°C, respectively, whereas the unsaturated lipids have their T_c below 0°C (see also Table XII). As a consequence of this fluidity most biological membranes can easily deform and bend as well as allow various solute molecules to pass through them.

A further important aspect of lipid chain fluidity is that different lipid types can pack together (i.e., mutually accommodate each other as well as other molecules) while remaining within a planar or curved bilayer configuration. In addition, the curvature of a lipid bilayer can be regulated by altering the ratio of its constituent lipids. For example, addition of lysolecithin to lecithin vesicles results in smaller vesicles since a mixture of such lipids can pack into more highly curved bilayers (cf. Table XXI). Only small amounts of lysolecithin can be incorporated into bilayers or biological membranes before they break up (become solubilized) into small vesicles or micelles. On the other hand, phosphatidylethanolamine and cholesterol both have the shape of an inverted truncated cone (Table XXI), and these when mixed with lecithin *increase* the critical radius of curvature (e.g., the vesicle size). Only a limited amount of cholesterol and phosphatidylethanolamine can be incorporated into bilayers before the bilayer structure is destroyed, at the point where the fluid lipids can no longer accommodate these inverted-cone lipids. Indeed, natural biological membranes never contain high amounts of both cholesterol and phosphatidylethanolamine. As might be expected, cholesterol and lysolecithin—neither of which can form a stable bilayer by itself—can combine in certain proportions to form bilayers. These and other physical properties of mixed lipid bilayers, vesicles, and biological membranes were discussed by Carnie *et al.* (1979) and Murphy (1982).

It is instructive to note that the most common lipids in animal cells are the two phospholipids phosphatidylcholine (PC) and phosphatidylethanolamine (PE), while in plant cells they are the glycolipids digalactosyl diglyceride (DGDG) and monogalactosyl diglyceride (MGDG). In each case the first lipid packs as a truncated cone ($v/a_0l_c < 1$) while the second packs as an inverted truncated cone ($v/a_0l_c > 1$). Thus, depending on the ratio of these lipid pairs in a bilayer, they can pack together into planar bilayers (Fig. 76a) or

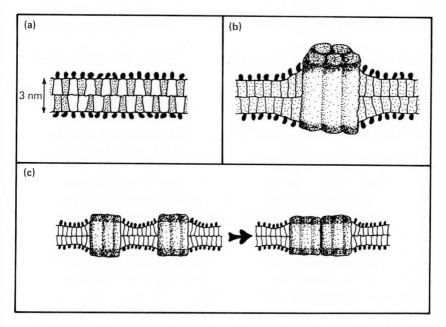

Fig. 76 (a) Mixture of two different lipids packing together within a planar membrane. One of the lipids has $v/a_0l_c < 1$ (cone shaped, such as lecithin or DGDG), the other has $v/a_0l_c > 1$ (wedge shaped, such as phosphatidyl ethanolamine, MGDG, or cholesterol).

(b) Packing constraints induced in the hydrocarbon chain regions of lipids around a protein molecule, which may be relaxed when proteins tilt (not shown) or aggregate, as shown in (c).

into bilayers of varying curvature and flexibility, a facility that is made use of by the lipid synthesis machinery of cells (Section 16.12). It is also worth noting that each of these lipids has an uncharged head-group. Their repulsive head-group interactions are due to hydration forces that are fairly insensitive to changes in the ionic environment of the cytoplasm. The invariant packing properties of PC, PE, DGDG, and MGDG make these lipids the ideal structural building blocks for stable membrane organization.

16.12. Membrane proteins and membrane structure

Membrane proteins are long-chained polypeptides—polymers consisting of a long string of amino acid residues. Their molecular weight can reach 100,000. Compared to hydrocarbon-chained lipids, membrane proteins are structurally rigid. The chains fold into a number of cylindrical α-helical segments, each exposing a hydrophobic surface, which are linked together by hydrophilic residues (Capaldi, 1982). The hydrophobic helices span the lipid bilayer into which they are incorporated, while the hydrophilic residues occupy the bilayer–water interfacial regions on either side of the membrane (Fig. 76b).

When proteins are incorporated into a lipid bilayer they usually affect the lipids in their vicinity. Such perturbed lipids are known as *boundary lipids*. What is the origin and nature of this perturbation? First, if the hydrophobic regions of the protein and bilayer have different lengths, then in the vicinity of a protein the lipid bilayer must distort in order to accommodate the protein. The fluid hydrocarbon chains of the lipids will distort while attempting to maintain the head-group area near the optimal value and at the same time not extending the chains beyond the maximum allowable length. Such packing perturbations around a protein usually involve more than one lipid layer as illustrated in (Fig. 76b). Additional lipid–protein interactions will include any specific electrostatic or hydrogen-bonding interactions between the hydrophilic head groups and the amino acid residues of proteins. Such lipid–protein interactions often result in a preferential clustering of specific lipids around a protein—the favoured lipids being those that can be packed most easily around the protein. For a recent review of lipid–protein interactions in membranes, see Benga and Holmes (1984).

Second, attractive or repulsive forces between proteins occur due to straightforward ion binding, but they may also arise as a consequence of a lipid–protein packing mismatch. Thus, the strain on the lipid packing may be relieved if proteins aggregate, as illustrated in Fig. 76c. As examples of this phenomenon, it has been found that the state of aggregation of rhodopsin and bacteriorhodopsin incorporated into phospholipid bilayers can be altered by

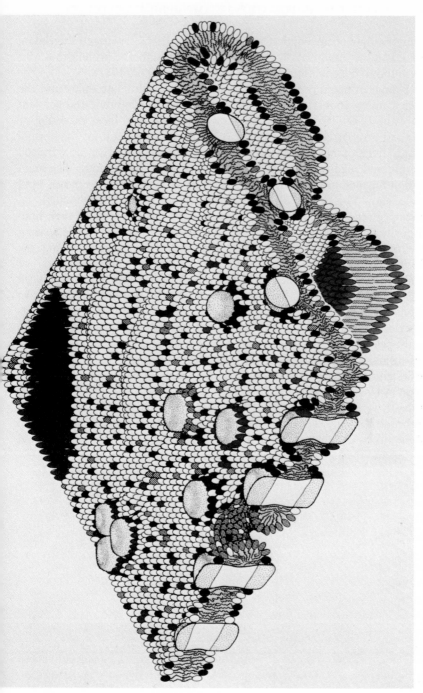

Fig. 77 Schematic figure of a biological membrane. The lipid contents of different membranes vary from as little as 25% up to 80% by weight. At highly curved regions the outer convex face contains mainly cone-shaped lipids while the inner concave face has more wedge-shaped lipids.

varying the lipid chain length (Chen and Hubbell, 1973; Lewis and Engelman, 1983a), and the organization of photosynthetic membrane proteins has also been found to depend on lipid composition (Siegel *et al.*, 1981).

A biological membrane is a dynamic structure. Both the lipids and the proteins move about rapidly in the plane of the membrane (Singer and Nicholson, 1972). However, heterogeneous domains and local clustering of lipids and proteins also occur (Fig. 77), which are important for the normal chemical functioning of the membrane and its components. How does the cell maintain and regulate the structural integrity of its membranes? The main criterion for membrane stability appears to be that the heterogeneous lipid mixture should be able to self-assemble into bilayers, even though individual species (e.g., cholesterol) may not. Thus, many organisms change their lipid composition in response to a change in ambient temperature, synthesizing more *unsaturated* lipids at *lower* temperature, thereby maintaining the stability of the bilayer structure, since the higher packing parameter of unsaturated lipids offsets the lowering of $v/a_0 l_c$ as the temperature is reduced (cf. Table XXI). Likewise, Wieslander *et al.* (1980) showed that the *in vivo* response of the bacterium *Acholeplasma laidlawii* to external stimuli, such as temperature changes and incorporation of fatty acids and cholesterol, is such that the cell synthesizes just those lipids that will compensate for these stimuli so as to maintain packing compatibility within the membrane, for example, synthesizing more DGDG ($v/a_0 l_c < 1$) and less MGDG ($v/a_0 l_c > 1$) after cholesterol is introduced.

These two examples illustrate the manner in which cells control their membrane structure. They do not do this by a crude mechanical pushing and pulling of components. The cell synthesizes the appropriate lipids and proteins and then leaves them to do their job.

Chapter 17

The Interactions between Lipid Bilayers and Biological Membranes

17.1. Attractive van der Waals and repulsive double-layer forces

In Part II we saw that there are four main types of forces acting between surfaces in aqueous solutions. These are the attractive *van der Waals* force, repulsive *electrostatic double-layer* force, attractive or repulsive *hydration* forces, and attractive or repulsive *steric* forces. In this final chapter we shall review these in turn, but now focussing our attention on their role in the interactions between fluid amphiphilic structures such as bilayers and membranes

Between two semi-infinite planar surfaces the van der Waals interaction free energy per unit area is given by

$$W(D) = -A/12\pi D^2, \qquad (17.1)$$

where A is the nonretarded *Hamaker constant* (Chapter 11). For interactions between hydrocarbons or lipid bilayers across aqueous solutions, A is relatively small, lying in the range $4-7 \times 10^{-21}$ J. Equation (17.1), when applied to bilayers or membranes, is only strictly valid at separations D below about 3 nm. Above 3 nm the Hamaker constant is no longer "constant" but diminishes progressively as D increases due to retardation effects and to the finite thickness of bilayers, and by 10 nm it is usually much less than half the short-distance (nonretarded) value. In the presence of electrolyte there is an additional reduction of A at finite separations due to the ionic screening of the zero-frequency contribution $A_{\nu=0}$, which is now given by (Mahanty and

Ninham, 1976, p. 202):

$$A_{v=0}(D) \simeq A_{v=0}(0)2\kappa D e^{-2\kappa D}, \qquad \text{for} \quad \kappa D \gg 1, \qquad (17.2)$$

where κ is the Debye length and where for hydrocarbon across water $A_{v=0}(0) = 3 \times 10^{-21}$ J (Section 11.4). Equation (17.2) is accurate to within 15% for $\kappa D > 2$.

In conclusion, the van der Waals force between bilayers and membranes in aqueous solutions is fairly weak and has an effective range of at most 15 nm, beyond which it is too weak to be of any major significance.

The double-layer repulsion, unlike the van der Waals attraction, is very sensitive to the type and concentration of electrolyte present, the pH, and the surface charge density (or potential). An approximate expression for the double-layer interaction energy between two surfaces is given by Eq. (12.47), while exact solutions are plotted in Fig. 46. As described in Chapter 12, the

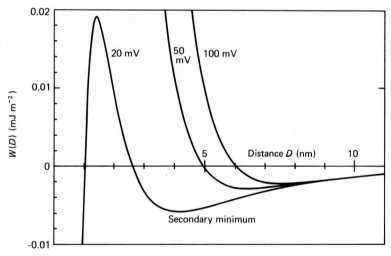

Fig. 78 DLVO interaction energy per unit area between two bilayers or membranes in 0.15 M NaCl (kindly computed by R. Pashley). The surfaces are assumed to interact at constant surface potential ψ_0, and the nonretarded Hamaker constant is assumed to be $A = 6 \times 10^{-21}$ J (lipid bilayers usually have a slightly lower value, while biological membranes a slightly higher value due to the presence of proteins whose refractive index is generally higher than that of fluid hydrocarbon chains, 1.55–1.60 compared to 1.42–1.48). The three curves at $\psi_0 = 100$, 50, and 20 mV correspond to surface charge densities of one electronic charge per 1, 3, and 8.5 nm², respectively. Note that the surfaces of real biological membranes are not smooth but locally rough due to protruding proteins (Fig. 77), which complicates the application of the DLVO theory at small distances. Further, below ~3 nm repulsive hydration forces may dominate the interaction depending on the hydrophilicity of the surface groups.

interplay between attractive van der Waals forces and repulsive double-layer forces forms the basis of the so called DLVO theory of colloid science, and both of these forces have now been directly measured between surfactant and lipid bilayers in aqueous solutions (Pashley and Israelachvili, 1981; Lis et al. 1982; Ohshima et al. , 1982; see also Fig. 41).

Because of the relatively weak van der Waals attraction between lipid bilayers in water, the double-layer repulsion is usually strong enough to keep their surfaces apart, even for low surface potentials and in high salt. For two bilayers in physiological saline solution (about 0.15 M NaCl), the secondary minimum occurs at 4–6 nm for typical values of ψ_0 and A, as shown in Fig. 78.

17.2. Hydration forces

The DLVO theory is based on the assumption that water may be treated as a continuum fluid, characterized solely by its bulk properties (e.g., its density and dielectric constant). At small surface separations this is no longer the case since the discrete molecular properties of water and the way water molecules interact specifically with surfaces give rise to a solvation or hydration force (Section 13.4). Repulsive hydration forces arise whenever water molecules hydrate (i.e., strongly bind to) hydrophilic surface groups (e.g., —OH, —PO_4^-, —$N(CH_3)_3^+$, —Na^+ and —Mg^{2+}) because of the energy needed to dehydrate these groups as two surfaces approach each other. Hydration forces have been measured between surfactant and lipid bilayers (see Figs. 41 and 54), and there is good evidence that they also occur between biological membranes (Rand et al., 1979; Duniec et al., 1981) as well as between DNA molecules in water (Rau et al., 1984).

The maximum range of interbilayer hydration forces so far measured is 1–3.5 nm, below which they rise steeply and (roughly) exponentially with a decay length of about 0.2–0.3 nm. Thus, they dominate over DLVO forces at small separations, preventing the coalescence in a primary minimum of bilayers, vesicles, and biological membranes. In particular, hydration forces are responsible for the lack of strong adhesion or aggregation of bilayers and vesicles composed of uncharged lipids such as lecithin and those possessing polyoxyethylene or sugar head groups (see Table XX).

In the case of negatively charged lipid vesicles (e.g., phosphatidylserine) the hydration force arising from Na^+ binding prevents the irreversible aggregation or fusion of these vesicles even in high salt (Day et al., 1980) where the DLVO theory would predict strong irreversible adhesion in molecular contact. Also, phosphatidylserine vesicles fuse in dilute (approximately millimolar) Ca^{2+} solutions but not in Mg^{2+} solutions (Wilschut et al., 1981), again because the more hydrated Mg^{2+} ions prevent the surfaces from coming into sufficiently close contact to fuse (McIver, 1979). As a further example of cation-regulated hydration forces, Princen *et al.* (1980) studied the interactions between oil-in-water emulsion droplets, whose surfaces were covered by an alkyl sulphate monolayer, and found that their adhesion in concentrated salt solutions decreases with the increasing hydration of the cation bound to the surfaces, namely KCl (maximum adhesion) > NaCl > LiCl (almost no adhesion). Other examples of the role of hydration in bilayer and membrane interactions were discussed by McIver (1979).

The origin of the monotonically repulsive hydration forces seen in so many different systems remains unclear. Marčelja's theory (Marčelja and Radić, 1976; Gruen and Marčelja, 1983) based on the polarization (alignment) induced by surfaces or surface groups on the water molecules adjacent to them appears to be the most likely explanation. Israelachvili and Pashley (1983) further suggested that any purely monotonic force profile may always have an oscillatory force superimposed on it. This was based on their measurements of hydration forces between rigid, molecularly smooth surfaces where such a force law was obtained (Fig. 57), and it is also in line with theoretical expectations of solvation forces in general. This being the case, one may well ask why only monotonically decaying forces have so far been measured between lipid bilayers. It may be that local fluctuations of bilayer thickness and the thermal motions of lipid head groups smear out any oscillatory component of the intrinsic hydration force, leaving only a monotonic background repulsion. For dipalmitoyl phosphatidylcholine, the spatial extent of head-group motion was measured by Büldt *et al.* (1979) who found that the first CH_2 group of the head group fluctuates normal to the bilayer surface with a Gaussian spread of a few angstroms both below and above T_c. Since the total length of lipid headgroups is quite substantial, usually greater than 1 nm, their segmental motions cannot be ignored when considering the interactions between them. The magnitude and range of the net hydration force between bilayers may thus involve the steric interactions of head groups (analogous to that of two surfaces with very short polymer chains adsorbed to them) as well as genuine solvation effects.

17.3. Other forces between amphiphilic surfaces

(i) The interaction between purely hydrophobic surfaces in water—the *hydrophobic* interaction—is attractive, and much stronger than the van der Waals attraction at small separations (see Sections 8.6 and 13.5). At present the interplay of the repulsive (hydration) and attractive (hydrophobic) solvation forces occurring between surfaces containing both hydrophilic and hydrophobic groups in close proximity is not known, but as discussed in Section 13.5

it appears that the two interactions are not additive and that the hydrophobic interaction may be neutralized in the presence of hydrophilic groups.

(ii)   The *steric* interactions (Section 13.6) between bilayers and membranes are still not well understood. As noted above, the interactions between lipid head groups may involve steric effects especially for long-chained head groups such as occur in polyoxyethylene surfactants (Table XX), which form micelles or bilayers in water whose interactions are very complex. Two other essentially steric-type interaction are worthy of note. The first involves polymer bridges (cf. Fig. 58f), which have been found to connect myelin membranes where their function is presumably to prevent these membranes from getting too far apart from each other (Rand *et al.*, 1979). The second concerns the interactions between the hydrocarbon chains of surfactant monolayers across hydrocarbon liquids. Such situations arise in water-in-oil microemulsions as well as when liquid alkanes are present in the central regions of bilayers. Gruen and Haydon (1981) studied this system, both experimentally and theoretically, and found that low molecular weight alkanes penetrate into the hydrocarbon chains leading to an effective repulsion between apposing bilayers while higher molecular weight alkanes (above decane) penetrate progressively less, and the interaction is now attractive. The van der Waals forces between hydrocarbon chains across hydrocarbon liquids is, of course, small and plays only a minor role in such systems.

(iii)   Finally, we mention *osmotic* forces, which may be particularly important in the interactions of biological membranes. Bilayers and membranes are highly permeable to water. This means that water will pass readily through a membrane when an osmotic gradient exists across it. Thus, if the aqueous space between two membranes or lipid vesicles has a lower salt concentration than the other aqueous compartments, water will flow away from the interspace between the membranes thereby creating an effective attractive force pulling the two membranes together. Such osmotic gradients have been found to be essential for triggering the fusion of certain bilayers and membranes (Zimmerberg *et al.*, 1980; Fisher and Parker, 1984).

17.4. Interdependence of intermembrane and intramembrane forces

Since the electrostatic, solvation, and steric interactions between adjacent head groups *within* a bilayer are the same as those occurring *between* two opposing bilayers, we should expect a correlation between intrabilayer forces and interbilayer forces (since both interactions involve the same head groups). Thus, we might expect larger head-group areas a_0 to be accompanied by larger

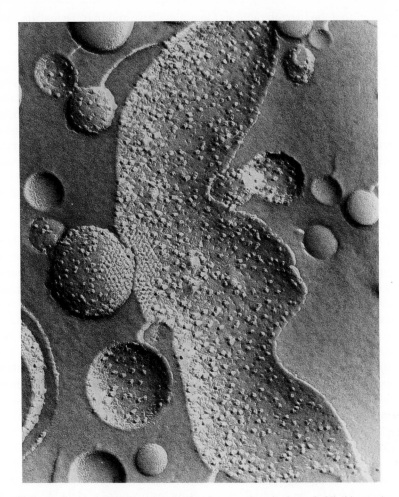

Fig. 79 Freeze-fracture electron micrograph of reconstituted plant thylakoid membranes containing plant lipids and the two major thylakoid proteins known as photosystem I (mean diameter 9 nm) and the light-harvesting complex of photosystem II (LHC-II, of mean diameter 7 nm). In the presence of 5 mM Mg^{2+} the LHC-II aggregate into two-dimensional crystalline patches. These act as adhesion sites between membranes, as seen here and also in intact thylakoids. The adhering vesicle in the figure has a diameter of 220 nm. (Micrograph: I. J. Ryrie.)

repulsions between bilayers. This is borne out by experiments. For example, the large hydration of the lecithin head group results in a large surface area of $a_0 \approx 0.7$ nm^2 as well as a large swelling in fully hydrated lecithin multilayers. By contrast the head-group repulsion in phosphatidylethanolamines is much less, which leads to a smaller head-group area of $a_0 \approx 0.45$–0.55 nm^2 and a much reduced swelling (viz., 1.3 nm compared to about 2.7 nm for lecithins). Analogous correlations occur in anionic lipid bilayers where in general a decrease in pH or addition of divalent cations reduces the electrostatic head-group repulsion and hence the surface area per lipid and also leads to reduced bilayer swelling in multilayer phases and to increased adhesion of vesicles.

As a further example, the aggregation number of alkyl sulphate micelles increases as the electrolyte is changed from $LiCl \rightarrow NaCl \rightarrow KCl \rightarrow CsCl$ (Missel et al., 1982). This implies a reduced head-group area a_0 (Chapter 16), which arises from the reduced hydration repulsion of the bound counterions as we go from Li^+ to Cs^+. This correlates with the already mentioned increasing adhesion between oil-in-water emulsion droplets, stabilized by alkyl sulphate monolayers, as monitored by the contact angles between them (Princen et al., 1980).

In the area of biological membrane interactions, many studies of membrane and vesicle fusion have found that proteins redistribute during the fusion process, and even in the absence of fusion both protein and lipid redistributions have been reported to occur at membrane–membrane contact sites. For example, the inner surface of the outer mitochondrial membrane (see Fig. 75) has a nonrandom distribution of anionic sites, with a lower density of sites in those areas surrounding the inner membrane, whereas the distribution of anionic sites on the exposed outer surface is random (Hackenbrock and Miller, 1975). In plant thylakoid membranes (Fig. 75) the random distribution of proteins characteristic of the isolated membranes becomes nonrandom on stacking where, moreover, the stacked (grana) regions contain different proteins from those in the unstacked (stroma) regions (Staehelin and Arntzen, 1979; Anderson and Andersson, 1982; Ryrie, 1983). Figure 79 illustrates this phenomenon, showing how two "reconstituted membranes" have adhered at the sites where specific proteins have aggregated.

17.5. Adhesion and fusion of vesicles and membranes

Compared to all the feverish goings on in a cell, the fusion of extended biological membranes does not occur regularly. However, the fusion of vesicles with certain specialized membranes is a common process, notably during synaptic nerve transmission, exocytosis (vesicle incorporation), and pinocytosis (vesicle shedding), as illustrated in Fig. 75. More recent interest in vesicles as potential drug carriers has also stimulated research into the way vesicles adhere and fuse with each other and with planar membranes (the Papahadjopoulos group at the University of California in San Francisco being particularly active in this area).

Pure lipid vesicles (e.g., phosphatidylcholine) in water are often stable for months but fuse rapidly (within minutes) into larger vesicles when the temperature is lowered much below the chain melting temperature T_c. This is because the highly curved bilayers of small vesicles become stressed below T_c as the frozen chains attempt to line up (i.e., to pack into less curved bilayers) (Table XXI). In this case rupture followed by fusion will occur during vesicle collisions, not because of any intervesicle interactions but because of changes in the intravesicle forces that have induced fragility in the vesicles.

Vesicle aggregation and fusion can also be brought about by changing the intervesicle forces. For example, negatively charged lipid vesicles of phosphatidylserine, phosphatidic acid, or phosphatidylglycerol aggregate then fuse in the presence of $1-10$ mM Ca^{2+} due to the reduced double-layer repulsion, an effect that is analogous to the coagulation of negatively charged colloidal particles in the presence of millimolar amounts of divalent cations (Chapter 12). It should be noted that here, too, even though the presence of Ca^{2+} increases the attraction between anionic vesicles, it also induces stresses in the lipid packing within vesicles in much the same way as lowering the temperature below T_c, an effect that further facilitates fusion. Anionic vesicles are less susceptible to fusion the more hydrated the bound cation and/or the more hydrated the head group. Thus, Mg^{2+} often induces reversible aggregation where Ca^{2+} induces fusion, and vesicles of phosphatidylinositol (which contains a bulky, highly hydrated head group) do not fuse in Ca^{2+} (or Mg^{2+}) solutions.

Calcium is usually involved in cellular fusion processes. But the fusion of biological vesicles or membranes, which contain calcium-specific proteins, can be triggered by much lower calcium concentrations, usually $1-10$ μM Ca^{2+}. Other biological membrane "fusogens" are lysolecithin, viruses, and osmotic gradients. And it is known that fusion occurs very rapidly and is usually over within 1 ms.

Clearly, the interactions of vesicles and membranes are complex, involving both inter- and intramembrane forces as well as specific interactions when proteins are also present. Let us try to analyse the simplest possible interaction between two vesicles in terms of what we already know about surface forces. The shape of the interaction potential between two membranes (Fig. 47) tells us whether they will attract or repel each other, and—if they attract—the depth of the potential or adhesive well, W. But there are two things it does not tell us. It does not, without further analysis, tell us how two vesicles will deform when they adhere, and in particular it gives no indication of whether fusion will occur nor, indeed, of the fusion mechanism itself. Let us consider these two phenomena in turn.

When a pure liquid droplet of surface tension γ settles on a flat surface (Fig. 80a) it distorts and becomes a truncated sphere. The contact angle θ is

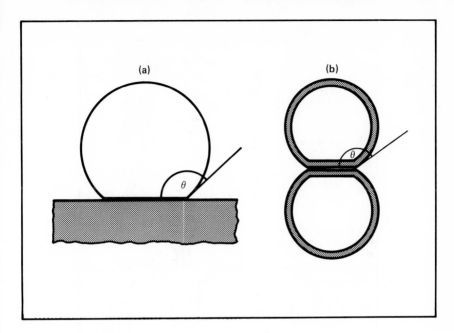

Fig. 80
(a) Liquid drop or large vesicle on surface.
(b) Two adhering vesicles.

given by the *Young–Dupré* equation (Section 14.2),

$$(1 + \cos\theta)\gamma = W,$$

where W is the adhesion energy (work of adhesion) per unit area of the two surfaces. In the case of a spherical vesicle approaching a surface, a similar deformation takes place on adhesive contact, but here the area elastic modulus k_c of the membrane determines θ. (Note that for a pure liquid surface, a change in its surface area by ΔA is accompanied by a surface free energy change of $\gamma \Delta A$, while for an elastic membrane this becomes replaced by $\frac{1}{2}k_c \Delta A^2/A_0$ as given by Eq. (16.13), where A_0 is the equilibrium, unstressed, area.) It is left as an exercise for the reader to establish that for a large spherical vesicle, initially unstressed, whose membrane stretches elastically with an elastic expansion modulus k_c, the equilibrium contact angle at constant vesicle volume (Fig. 80a) is given by

$$(1 + \cos\theta)k_c = W \left/ \left\{ \frac{(3 + \cos\theta)}{[2(1 - \cos\theta)^{1/2}(2 + \cos\theta)]^{2/3}} - 1 \right\}, \right. \qquad (17.3)$$

where W is the adhesion energy per unit area of the two surfaces as before. For

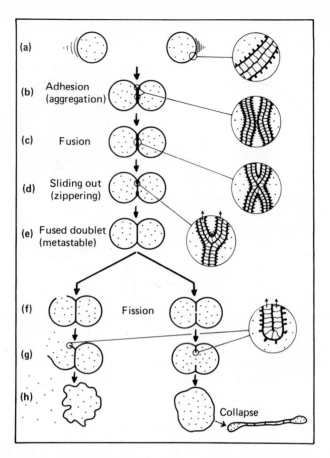

Fig. 81   Probable molecular events taking place during the adhesion, fusion, and fission of bilayers and membranes.

(a)   *Approach* of two vesicles.

(b)   *Adhesion* (aggregation): bilayers in tension with consequent exposure of more hydrophobic surface regions to each other. Rupture or local breakthrough may now occur, leading to the following.

(c)–(e)   *Fusion*: via the rapid sliding out of the outer monolayer lipids (zippering fusion). The final stage (e) can be a long-lived metastable state. Note that so far the vesicles' contents remain "internalized."

(f)–(h)   *Fission* (right side): here the final vesicle (h) has all the original outer monolayer lipids on the outside and all the inner monolayer lipids on the inside. The vesicles' contents remain "internalized."

  Fission (left side): Alternative fission pathway. Not all the vesicles' contents are "internalized." Here both the outer and inner monolayer lipids of one of the vesicles go to the outer monolayer of the final vesicle, resulting in high stresses in the membrane that would have to be alleviated somehow (e.g. via flip-flop). For this reason this fission process is energetically unfavourable compared to the one on the right.

two identical vesicles (Fig. 80b), k_c becomes replaced by $2k_c$. Further, it may be established that for two identical vesicles of initial radius R,

(i)   the total adhesion energy at contact equilibrium is proportional to R^2 rather than to R as for rigid (underformable) spheres,

(ii)   the adhesion force is $2\pi RW$, and

(iii)   for typical values of $k_c \approx 100$ mJ m^{-2}, the contact angle θ will fall below $\sim 135°$ once the work of adhesion W exceeds 1 mJ m^{-2}. Such high values of W are expected only when repulsive hydration and double-layer forces are suppressed. In such cases the total surface area of each vesicle will be stretched by $> 2\%$ of its initial (unstressed) value. Since most lipid bilayers and biological membranes cannot be stretched beyond 2–4% without rupturing (Wobschall, 1971; Evans and Skalak, 1980), this would lead to rupture, followed by fusion of vesicles.

In conclusion, vesicle and membrane fusion may result from a number of factors. First, a vesicle bilayer may be internally stressed or destabilized (e.g., by changes in the ionic environment or temperature). If such changes favour larger vesicles (due to an altered packing of lipids), they may lead to rupture followed by the association of ruptured vesicles. In such cases fusion may not be due to intervesicle forces *per se*. Second, vesicles may collide with each other and/or adhere and then rupture due to the increase in their surface areas on collision or adhesion. Again this would result in fusion, but here the fusion will also depend on the strength of the attractive forces (or the reduction in the repulsive forces) between vesicles. It is clear that more specific binding mechanisms are involved in the fusion of biological membranes, though there are indications (Horn, 1984) that the fusion processes of both lipid and biological membrane vesicles proceed via the same three intermediate stages, aggregation (adhesion) → fusion → fission, as shown in Fig. 81.

References

Abraham, F. F. (1978). *J. Chem. Phys.* **68**, 3713–3716.

Adam, N. K., and Stevenson, D. G. (1953). *Endeavour* **XII**, (45), 25–32.

Adamson, A. W. (1976). "Physical Chemistry of Surfaces," 3rd ed., Wiley, New York and London.

Alder, B. J., Hoover, H. G., and Young, D. A. (1968). *J. Chem. Phys.* **49**, 3688–3696.

Alexander, A. E., and Johnson, P. (1950). "Colloid Science," ch. XXIV, Oxford Univ. Press (Clarendon), London and New York.

Amis, E. S. (1975). *In* "Solutions and solubilities" (M. R. J. Dack, ed.), Part 1, pp. 105–193. Wiley (Interscience), New York and London.

Anderson, J. M., and Andersson, B. (1982). *Trends Biochem. Sci.* **7**, 288–292.

Aniansson, E. A. G., Wall, S. N., Almgren, M., Hoffmann, H., Kielmann, I., Ulbricht, W., Zana, R., Lang, J., and Tondre, C. (1976). *J. Phys. Chem.* **80**, 905–922.

Aveyard, R., and Saleem, S. M. (1976). *J. Chem. Soc. Faraday Trans. 1* **72**, 1609–1617.

Bailey, A. I., Price, A. G., and Kay, S. M. (1970). *Spec. Discuss. Faraday Soc.* **1**, 118–127.

Benga, G., and Holmes, R. P. (1984). *Prog. Biophys. Molec. Biol.* **43**, 195–257.

Ben Naim, A., Wilf, J., and Yaacobi, M. (1973). *J. Phys. Chem.* **77**, 95–102.

Blake, T. D. (1975). *J. Chem. Soc. Faraday Trans. 1* **71**, 192–208.

Bleaney, B. I., and Bleaney, B. (1959). "Electricity and magnetism," Oxford Univ. Press (Clarendon), London and New York.

Bloomquist, C. R., and Shutt, R. S. (1940). *Ind. Eng. Chem.* **32**(6), 827–831.

Bockris, J. O'M., and Reddy, A. K. N. (1970). "Modern Electrochemistry," vol. 1, ch. 2, Plenum, New York.

Bondi, A. (1968). "Physical properties of molecular crystals, liquids, and glasses," Wiley, New York and London.

Bradley, R. S. (1932). *Phil. Mag.* **13**, 853–862.

Brunner, J., Skrabal, P., and Hauser, H. (1976). *Biochim. Biophys. Acta* **455**, 322–331.

Büldt, G., Gally, H. U., Seelig, J., and Zaccai, G. (1979). *J. Mol. Biol.* **134**, 673–691.

Cain, F. W., Ottewill, R. H., and Smitham, J. B. (1978). *Faraday Discuss. Chem. Soc.* **65**, 33–42.

Capaldi, R. A. (1982). *Trends Biochem. Sci.* **7**, 292–295.

Carnie, S. L., and Chan, D. Y. C. (1982). *Adv. Colloid Interface Sci.* **16**, 81–100.

Carnie, S., Israelachvili, J. N., and Pailthorpe, B. A. (1979). *Biochim. Biophys. Acta* **554**, 340–357.

Castellan, G. W. (1972). "Physical Chemistry," 2nd ed., Addison-Wesley, Reading Massachusetts.

Chan, D. Y. C., Mitchell, D. J., Ninham, B. W., and Pailthorpe, B. A. (1980a). *J. Chem. Soc. Faraday Trans. 2* **76**, 776–784.

Chan, D. Y. C., Pashley, R. M., and White, L. R. (1980b). *J. Colloid Interface Sci.* **77**, 283–285.

Chen, Y. S., and Hubbell, W. L. (1973). *Exp. Eye Res.* **17**, 517–532.

Christenson, H. K. (1983). Ph.D. Thesis, Australian National University, Canberra, Australia. Also *J. Chem. Phys.* **78**, 6906–6913.

Christenson, H. K., and Horn, R. G. (1983). *Chem. Phys. Lett.* **98**, 45–48.

Christou, N. I., Whitehouse, J. S., Nicholson, D., and Parsonage, N. G. (1981). *Symp. Faraday Soc.* **16**, 139–149.

Clark, V. A., Tittmann, B. R., and Spencer, T. W. (1980). *J. Geophys. Res.* **85**, 5190–5198.

Clunie, J. S., Goodman, J. F., and Symons, P. C. (1967). *Nature* **216**, 1203–1204.

Coakley, C. J., and Tabor, D., (1978). *J. Phys. D* **11**, L77–L82.

Cotton, F. A., and Wilkinson, G. (1980). "Advanced Inorganic Chemistry," 4th ed., p. 1188, Wiley, New York.

Coulson, C. A. (1961). "Valence," 2nd ed., ch. XIII, Oxford Univ. Press, London and New York.

Cowley, A. C., Fuller, N. L., Rand, R. P., and Parsegian, V. A. (1978). *Biochemistry* **17**, 3163–3168.

Croxton, C. A. (1975). "Introduction to Liquid State Physics," Wiley, New York.

Dasent, W. E. (1970). "Inorganic Energetics," Penguin Books, U. K., U.S.A., and Australia.

Dashevsky, V. G., and Sarkisov, G. N. (1974). *Mol. Phys.* **27**, 1271–1290.

Davies, M. (1965). "Some Electrical and Optical Aspects of Molecular Behaviour," Pergamon, Oxford.

Day, E. P., Kwok, A. Y. W., Hark, S. K., Ho, J. T., Vail, W. J., Benz, J., and Nir, S. (1980). *Proc. Natl. Acad. Sci. USA* **77,** 4026–4029.

de Boer, J. H. (1936). *Trans. Faraday Soc.* **32,** 10–38.

Del Pennino, U., Mazzega, E., Valeri, S., Alietti, A., Brigatti, M. F., and Poppi, L. (1981). *J. Colloid Interface Sci.* **84,** 301–309.

Denbigh, K. G. (1940). *Trans. Faraday Soc.* **36,** 936–948.

Derjaguin, B. V. (1934). *Kolloid Zeits.* **69,** 155–164.

Derjaguin, B. V., and Abrikossova, I. I. (1954). *Discuss. Faraday Soc.* **18,** 24–41.

Derjaguin, B. V., and Churaev, N. V. (1974). *J. Colloid Interface Sci.* **49,** 249–255.

Derjaguin, B. V., and Churaev, N. V. (1977). *Croat. Chem. Acta* **50,** 187–195.

Derjaguin, B. V., and Kusakov, M. M. (1939). *Acta Physicochim. URSS* **10,** 25–44, 153–174.

Derjaguin, B. V., and Landau, L. (1941). *Acta Physicochim. URSS* **14,** 633–662.

Derjaguin, B. V., and Titijevskaia, A. S. (1954). *Discuss. Faraday Soc.* **18,** 24–41.

Derjaguin, B. V., Muller, V. M., and Toporov, Y. P. (1975). *J. Colloid Interface Sci.* **53,** 314–326.

Derjaguin, B. V., Rabinovich, Y. I., and Churaev, N. V. (1978). *Nature* **272,** 313–318.

Donners, W. A. B., Rijnbout, J. B., and Vrij, J. (1977). *J. Colloid Interface Sci.* **61,** 249–260.

Duniec, J. T., Israelachvili, J. N., Ninham, B. W., Pashley, R. M., and Thorne, S. W. (1981). *FEBS Lett.* **129,** 193–196.

Dymond, J. H. (1981). *J. Phys. Chem.* **85,** 3291–3294.

Dzyaloshinskii, I. E., Lifshitz, E. M., and Pitaevskii, L. P. (1961). *Adv. Phys.* **10,** 165–209.

Eagland, D. (1975). *In* "Water: A Comprehensive Treatise" (F. Franks, ed.), vol. 5, ch. 1, Plenum, New York.

Ekwall, P. (1975). *Adv. Liq. Cryst.* **1,** 1–142.

Eicke, H. F. (1980). *Top. Curr. Chem.* **87,** 85–145.

Evans, E. A., and Skalak, R. (1980). "Mechanics and Thermodynamics of Biomembranes," CRC Press, Boca Raton, Florida.

Evans, D. F., Tominaga, T., and Davis, H. T. (1981). *J. Chem. Phys.* **74,** 1298–1305.

Finean, J. B., Coleman, R., and Mitchell, R. H. (1978). "Membranes and their Cellular Functions," 2nd ed., Blackwell Scientific Publications, Oxford.

Finney, J. L. (1982). *In* "Biophysics of Water" (F. Franks, ed.), pp. 73–95, Wiley, New York.

Fisher, L. R., and Israelachvili, J. N. (1981). *Colloids Surf.* **3,** 303–319.

Fisher, L. R., and Parker, N. S. (1984). *Biophys. J.* **46**, 253–258.

Flory, P. J. (1953). "Principles of Polymer Chemistry," Cornell University Press, Ithaca, New York.

Forsyth, P. A., Marčelja, S., Mitchell, D. J., and Ninham, B. W. (1977). *Biochim. Biophys. Acta* **469**, 335–344.

Forsyth, P. A., Marčelja, S., Mitchell, D. J., and Ninham, B. W. (1978). *Adv. Colloid Interface Sci.* **9**, 37–60.

Fowkes, F. F. (1964). *Ind. Eng. Chem.* **56**(12), 40–52.

Franks, F. (1972–1982). "Water: A comprehensive treatise" (F. Franks, ed.), vols. 1–7, Plenum, New York and London.

Gallot, B., and Skoulios, A. (1966). *Kolloid Z.U.Z. Polymere* **208**, 37–43.

Georges, J. M. (ed.) (1982). "Microscopic aspects of adhesion and lubrication," Elsevier, Amsterdam.

Gingell, D., and Parsegian, V. A. (1972). *J. Theor. Biol.* **36**, 41–52.

Good, R. J., and Elbing, E., (1970). *Ind. Eng. Chem.* **62**(3), 54–78.

Grahame, D. C. (1953). *J. Chem. Phys.* **21**, 1054–1060.

Gregory, J. (1970). *Adv. Colloid Interface Sci.* **2**, 396–417.

Gregory, J. (1973). *J. Chem. Soc. Faraday Trans. 2* **69**, 1723–1728.

Gregory, J. (1975). *J. Colloid Interface Sci.* **51**, 44–51.

Gregory, J. (1978). *In* "The Scientific Basis of Flocculation" (K. J. Ives, ed.) pp. 101–130, NATO Adv. Study Inst. Series E., No. 27, Sijthoff & Noordhoff, The Netherlands.

Grimson, M. J., Rickayzen, G., and Richmond, P. (1980). *Mol. Phys.* **39**, 61–73, 1455–1462.

Gruen, D. W. R. (1981). *J. Colloid Interface Sci.* **84**, 281–283.

Gruen, D. W. R., and de Lacey, E. H. B. (1984). *In* "Surfactants in Solution" (K. L. Mittal and B. Lindman, eds.), vol. 1, pp. 279–306, Plenum, New York.

Gruen, D. W. R., and Haydon, D. A. (1981). *Biophys. J.* **33**, 167–188.

Gruen, D. W. R., and Marčelja, S. (1983). *J. Chem. Soc. Faraday Trans. 2* **79**, 225–242.

Guggenheim, E. A. (1949). "Thermodynamics," North-Holland Publ., Amsterdam.

Hackenbrock, C. R., and Miller, K. J. (1975). *J. Cell. Biol.* **65**, 615–630.

Hall, D. G., and Pethica, B. A. (1967). *In* "Nonionic Surfactants" (M. J. Schick, ed.), ch. 16, Marcel Dekker, New York.

Hamaker, H. C. (1937). *Physica* **4**, 1058–1072.

Hamnerius, Y., Lundström, I., Paulsson, L. E., Fontell, K., and Wennerström, H. (1978). *Chem. Phys. Lipids* **22**, 135–140.

Harrison, R., and Lunt, G. G. (1980). "Biological Membranes," 2nd ed., Halsted Press, Wiley, New York.

Hauser, H. (1984). *Biochim. Biophys. Acta* **772**, 37–50.

Hauser, H., and Gains, N. (1982). *Proc. Natl. Acad. Sci. USA* **79**, 1683–1687.

Healy, T. W., and White, L. R. (1978). *Adv. Colloid Interface Sci.* **9**, 303–345.

Healy, T. W., Homola, A., and James, R. O. (1978). *Faraday Discuss. Chem. Soc.* **65**, 156–163.

Healy, T. W., Chan, D., and White, L. R. (1980). *Pure Appl. Chem.* **52**, 1207–1219.

Henry, J. D. Jr., Prudich, M. E., and Chak, L. (1980). *Colloids Surf.* **1**, 335–348.

Hermann, R. B. (1972). *J. Phys. Chem.* **76**, 2754–2759.

Hermann, R. B. (1975). *J. Phys. Chem.* **79**, 163–169.

Hertz, H. G. (1973). In "Water: A comprehensive treatise" (F. Franks, ed.), vol. 3, ch. 7, Plenum, New York and London.

Hesselink, F. Th. (1971). *J. Phys. Chem.* **75**, 65–71.

Hesselink, F. Th., Vrij, A., and Overbeek, J. Th. G. (1971). *J. Phys. Chem.* **75**, 2094–2103.

Hiemenz, P. C. (1977). "Principles of Colloid and Surface Chemistry," Dekker, New York and Basel.

Hill, T. L. (1960). "Introduction to statistical thermodynamics," ch. 2, Addison-Wesley, Reading, Massachusetts.

Hill, T. L. (1963, 1964). "Thermodynamics of Small Systems." Benjamin, New York.

Hirschfelder, J. O., Curtiss, C. F., and Bird, R. B. (1954). "Molecular Theory of Gases and Liquids," Wiley, New York; Chapman and Hall, London.

Hogg, R., Healy, T. W., and Fuerstenau, D. W. (1966). *Trans. Faraday Soc.* **62**, 1638–1651.

Honig, E. P., and Mul, P. M. (1971). *J. Colloid Interface Sci.* **36**, 258–272.

Horn, R. G. (1984). *Biochim. Biophys. Acta* **778**, 224–228.

Horn, R. G., and Israelachvili, J. N. (1981). *J. Chem. Phys.* **75**, 1400–1411.

Hough, D. B., and White, L. R. (1980). *Adv. Colloid Interface Sci.* **14**, 3–41.

Israelachvili, J. N. (1972). *Proc. Roy. Soc. Lond. A* **331**, 39–55.

Israelachvili, J. N. (1973). *J. Chem. Soc. Faraday Trans. 2* **69**, 1729–1738.

Israelachvili, J. N. (1974). *Q. Rev. Biophys.* **6**, 341–387.

Israelachvili, J. N. (1982). *Adv. Colloid Interface Sci.* **16**, 31–47.

Israelachvili, J. N., and Adams, G. E. (1978). *J. Chem. Soc. Faraday Trans. 1* **74**, 975–1001.

Israelachvili, J. N., and Ninham, B. W. (1977). *J. Colloid Interface Sci.* **58**, 14–25.

Israelachvili, J. N., and Pashley, R. M. (1982a). In "Biophysics of Water" (F. Franks, ed.), pp. 183–194, Wiley, New York.

Israelachvili, J. N., and Pashley, R. M. (1982b). *Nature* **300**, 341–342.

Israelachvili, J. N., and Pashley, R. M. (1983). *Nature* **306**, 249–250.

Israelachvili, J. N., and Tabor, D. (1972). *Proc. Roy. Soc. Lond. A* **331**, 19–38.

Israelachvili, J. N., and Tabor, D. (1973). *Prog. Surf. Membr. Sci.* **7**, 1–55.

Israelachvili, J. N., Mitchell, D. J., and Ninham, B. W. (1976). *J. Chem. Soc. Faraday Trans. I* **72**, 1525–1568.

Israelachvili, J. N., Mitchell, D. J., and Ninham, R. W. (1977). *Biochim. Biophys. Acta* **470**, 185–201.

Israelachvili, J. N., Tandon, R. K., and White, L. R. (1979). *Nature* **277**, 120–121.

Israelachvili, J. N., Marčelja, S., and Horn, R. G. (1980a). *Q. Rev. Biophys.* **13**(2), 121–200.

Israelachvili, J. N., Perez, E., and Tandon, R. K. (1980b). *J. Colloid Interface Sci.* **78**, 260–261.

Israelachvili, J. N., Tandon, R. K., and White, L. R. (1980c). *J. Colloid Interface Sci.* **78**, 430–443.

Israelachvili, J. N., Tirrell, M., Klein, J. and Almog, Y. (1984). *Macromolecules* **17**, 204–209.

Joesten, M. D., and Schaad, L. J. (1974). "Hydrogen Bonding," Dekker, New York.

Johnson, K. L., Kendall, K., and Roberts, A. D. (1971). *Proc. Roy. Soc. London, Ser. A* **324**, 301–313.

Jönsson, B. (1981). *Chem. Phys. Lett.* **82**, 520–525.

Jönsson, B., and Wennerström, H. (1981). *J. Colloid Interface Sci.* **80**, 482–496.

Jönsson, B., and Wennerström, H. (1983). *J. Chem. Soc. Faraday Trans. 2* **79**, 19–35.

Jönsson, B., Wennerström, H., and Halle, B. (1980). *J. Phys. Chem.* **84**, 2179–2185.

Kauzmann, W. (1959). *Adv. Protein Chem.* **14**, 1–63.

Klein, J. (1980). *Nature* **288**, 248–250.

Klein, J. (1982). *Adv. Colloid Interface Sci.* **16**, 101–115.

Klein, J., and Luckham, P. (1982). *Nature* **300**, 429–431.

Klein, J., and Luckham, P. (1984). *Nature* **308**, 836–837.

Kohler, F. (1972). "The Liquid State," Verlag Chemie, Weinheim.

Kruus, P. (1977). "Liquids and Solutions: Structure and Dynamics," Dekker, New York.

Kumar, R., and Prausnitz, J. M. (1975). *In* "Solutions and Solubilities" (M. R. J. Dack, ed.), part 1, pp. 259–326, Wiley (Interscience) New York.

Landau, L. D., and Lifshitz, E. M. (1963). "Electrodynamics of Continuous Media," vol. 8, 2nd ed., Pergamon, Oxford.

Landau, L. D., and Lifshitz, E. M. (1980). "Statistical Physics," 2nd ed., part 1, Pergamon, Oxford.

Langmuir, I. (1938). *Science* **88**, 430–432.

Laughlin, R. G. (1978). *In* "Advances in Liquid Crystals" (G. H. Brown, ed.), vol. 3, pp. 41–148, Academic Press, New York.

Laughlin, R. G. (1981). *J. Soc. Cosmet. Chem.* **32,** 371–392.

Lennard-Jones, J. E., and Dent, B. M. (1928). *Trans. Faraday Soc.* **24,** 92–108.

Lewis, B. A., and Engelman, D. M. (1983a). *J. Mol. Biol.* **166,** 203–210.

Lewis, B. A., and Engelman, D. M. (1983b). *J. Mol. Biol.* **166,** 211–217.

Lifshitz, E. M. (1956). *Soviet Physics JETP (Engl. Transl.)* **2,** 73–83.

Lighthill, M. J. (1970). "Introduction to Fourier Analysis and Generalized Functions," Cambridge Univ. Press, London and New York.

Lindblom, G., and Wennerström, H. (1977). *Biophys. Chem.* **6,** 167–171.

Lis, L. J., McAlister, M., Fuller, N., Rand, R. P., and Parsegian, V. A. (1982). *Biophys. J.* **37,** 657–665.

London, F. (1937). *Trans. Faraday Soc.* **33,** 8–26.

Lyklema, J., and Mysels, K. J. (1965). *J. Am. Chem. Soc.* **87,** 2539–2546.

Lyklema, J., and van Vliet, T. (1978). *Faraday Discuss. Chem. Soc.* **65,** 25–32.

Lyons, J. S., Furlong, D. N., and Healy, T. W. (1981). *Aust. J. Chem.* **34,** 1177–1187.

Mahanty, J., and Ninham, B. W. (1976). "Dispersion Forces," Academic Press, New York.

Maitland, G., Rigby, M., Smith, E., and Wakeham, W. (1981). "Intermolecular Forces: Their Origin and Determination," Oxford Univ. Press, New York.

Marčelja, S. (1973). *Nature* **241,** 451–453.

Marčelja, S. (1974). *Biochim. Biophys. Acta* **367,** 165–176.

Marčelja, S., and Radić, N. (1976). *Chem. Phys. Lett.* **42,** 129–130.

Marčelja, S., Mitchell, D. J., Ninham, B. W., and Sculley, M. J. (1977). *J. Chem. Soc. Faraday Trans. 2* **73,** 630–648.

Margenau, H., and Kestner, N. R. (1971). "Theory of intermolecular forces," Pergamon, Oxford.

Mazer, N. A., Benedek, G. B., and Carey, M. C. (1976). *J. Phys. Chem.* **80,** 1075–1085.

McFarlane, J. S., and Tabor, D. (1950). *Proc. Roy. Soc. London, Ser. A* **202,** 224–243.

McIver, D. J. L. (1979). *Physiol. Chem. Phys.* **11,** 289–302.

McLachlan, A. D. (1963a). *Proc. Roy. Soc. London, Ser. A* **271,** 387–401; **274,** 80–90.

McLachlan, A. D. (1963b). *Mol. Phys.* **6,** 423–427.

McLachlan, A. D. (1965). *Discuss. Faraday Soc.* **40,** 239–245.

McLaughlin, S., Mulrine, N., Gresalfi, T., Vaio, G., and McLaughlin, A. (1981). *J. Gen. Physiol.* **77,** 445–473.

Missel, P. J., Mazer, N. A., Carey, M. C., and Benedek, G. B. (1982). *In* "Solution Behaviour of Surfactants" (K. L. Mittal and E. J. Fendler, eds.), vol. 1, pp. 373–388, Plenum, New York.

Missel, P. J., Mazer, N. A., Benedek, G. B., and Carey, M. C. (1983). *J. Phys. Chem.* **87,** 1264–1277.

Moelwyn-Hughes, E. A. (1961). "Physical Chemistry," 2nd ed., Pergamon, Oxford.

Moore, W. J. (1974). "Physical Chemistry," ch. 7, Longman, London.

Muller, V. M., Yushchenko, V. S., and Derjaguin, B. V. (1980). *J. Colloid Interface Sci.* **77**, 91–101.

Muller, V. M., Derjaguin, B. V., and Toporov, Y. P. (1983). *Colloids Surf.* **7**, 251–259.

Murphy, D. J. (1982). *FEBS Lett.* **150**, 19–26.

Napper, D. H. (1977). *J. Colloid Interface Sci.* **58**, 390–407.

Nicholson, D., and Parsonage, N. G. (1982). "Computer Simulation and the Statistical Mechanics of Adsorption," ch. 8, Academic Press, New York.

Nightingale, E. R. (1959). *J. Phys. Chem.* **63**, 1381–1387.

Ninham, B. W., and Parsegian, V. A. (1970). *Biophys. J.* **10**, 646–663.

Ninham, B. W., and Parsegian, V. A. (1971). *J. Theor. Biol.* **31**, 405–428.

Norrish, K. (1954). *Discuss. Faraday Soc.* **18**, 120–134.

Ohshima, H., Inoko, Y., and Mitsui, T. (1982). *J. Colloid Interface Sci.* **86**, 57–72.

Onsager, L. (1936). *J. Am. Chem. Soc.* **58**, 1486–1493.

Orr, F. M., Scriven, L. E., and Rivas, A. P. (1975). *J. Fluid Mech.* **67**, 723–742.

Pangali, C., Rao, M., and Berne, B. J. (1979). *J. Chem. Phys.* **71**, 2975–2981.

Parfitt, G. D., and Peacock, J. (1978). *Surf. Colloid Sci.* **10**, 163–226.

Parsegian, V. A. (1966). *Trans. Faraday Soc.* **62**, 848–860.

Parsegian, V. A. (1973). *Annu. Rev. Biophys. Bioeng.* **2**, 221–255.

Parsegian, V. A., and Gingell, D. (1972). *Biophys. J.* **12**, 1192–1204.

Parsegian, V. A., and Weiss, G. H. (1981). *J. Colloid Interface Sci.* **81**, 285–289.

Parsegian, V. A., Fuller, N., and Rand, R. P. (1979). *Proc. Natl. Acad. Sci. USA* **76**, 2750–2754.

Parsons, J. M., Siska, P. E., and Lee, Y. T. (1972). *J. Chem. Phys.* **56**, 1511–1515.

Pashley, R. M. (1977). *J. Colloid Interface Sci.* **62**, 344–347.

Pashley, R. M. (1980). *J. Colloid Interface Sci.* **78**, 246–248.

Pashley, R. M. (1981a). *J. Colloid Interface Sci.* **80**, 153–162.

Pashley, R. M. (1981b). *J. Colloid Interface Sci.* **83**, 531–545.

Pashley, R. M., and Israelachvili, J. N. (1981). *Colloids Surf.* **2**, 169–187.

Pashley, R. M., and Israelachvili, J. N. (1984). *J. Colloid Interface Sci.* **97**, 446–455.

Pashley, R. M., and Kitchener, J. A. (1979). *J. Colloid Interface Sci.* **71**, 491–500.

Pass, G. (1973). "Ions in Solution 3: Inorganic Properties," Oxford Chemical Series, Oxford Univ. Press (Clarendon), London and New York.

Paterson, M. S., and Kekulawala, K. R. S. S. (1979). *Bull. Mineral.* **102**, 92–98.

Pauling, L. (1960). "The Nature of the Chemical Bond," 3rd ed., Cornell University Press, Ithaca, New York.

Payens, Th. A. J. (1955). *Philips Res. Rep.* **10**, 425–481.

Peschel, G., Belouschek, P., Müller, M. M., Müller, M. R., and König, R. (1982). *Colloid Polym. Sci.* **260**, 444–451.

Pratt, L. R., and Chandler, D. (1977). *J. Chem. Phys.* **67**, 3683–3704.

Princen, H. M., Aronson, M. P., and Moser, J. C. (1980). *J. Colloid Interface Sci.* **75**, 246–270.

Pryde, J. A. (1966). "The Liquid State," Hutchinson Univ. Library, London.

Quinn, P. J. (1976). "The Molecular Biology of Cell Membranes," Macmillan Press, London and Basingstoke.

Rabinovich, Y. I., Derjaguin, B. V., and Churaev, N. V. (1982). *Adv. Colloid Interface Sci.* **16**, 63–78.

Rand, R. P., Fuller, N. L., and Lis, L. J. (1979). *Nature* **279**, 258–260.

Rao, M., Berne, B. J., Percus, J. K., and Kalos, M. H. (1979). *J. Chem. Phys.* **71**, 3802–3806.

Rau, D. R., Lee, B. K., and Parsegian, V. A. (1984). *Proc. Natl. Acad. Sci. USA* **81**, 2621–2625.

Read, A. D., and Kitchener, J. A. (1969). *J. Colloid Interface Sci.* **30**, 391–398.

Reiss-Husson, F. (1967). *J. Mol. Biol.* **25**, 363–382.

Requena, J., Billett, D. F., and Haydon, D. A. (1975). *Proc. Roy. Soc. London A* **347**, 141–159.

Rickayzen, G., and Richmond, P. (1985). *In* "Thin Liquid Films" (I. B. Ivanov, ed.), ch. IV, Dekker, New York.

Robertson, R. N. (1983). "The Lively Membranes," Cambridge Univ. Press, London and New York.

Rushbrooke, G. S. (1940). *Trans. Faraday Soc.* **36**, 1055–1062.

Ryrie, I. J. (1983). *Eur. J. Biochem.* **137**, 205–213.

Sabisky, E. S., and Anderson, C. H. (1973). *Phys. Rev. A.* **7**, 790–806.

Saluja, P. P. S. (1976). "Int. Rev. Sci. Electrochemistry" (A. D. Buckingham, ed.), ch. 1, pp. 1–51, *Phys. Chem. Series 2*, vol. 6, Butterworth, London.

Savkoor, A. R. (1982). *In* "Microscopic aspects of adhesion and lubrication" (J. M. Georges, ed.), pp. 279–308, Elsevier, Amsterdam.

Scheutjens, J. M. H. M., and Fleer, G. J. (1982). *Adv. Colloid Interface Sci.* **16**, 361–380.

Schuster, P., Zundel, G., and Sandorfy, C. (1976). "The Hydrogen Bond," vols. I,II,III, North-Holland Publ., Amsterdam.

Shaw, D. J. (1970). "Introduction to Colloid and Surface Chemistry," Butterworth, London.

Shchukin, E. D. (1982). *In* "Microscopic Aspects of Adhesion and Lubrication" (J. M. Georges, ed.), pp. 389–402, Elsevier, Amsterdam.

Shchukin, E. D., Amelina, E. A., and Yaminsky, V. V. (1981). *Colloids Surf.* **2**, 221–242.

Shinitzky, M., Dianoux, A. C., Gitler, C., and Weber, G. (1971). *Biochemistry* **10**, 2106–2113.

Siegel, C. O., Jordan, A. E., and Miller, K. R. (1981). *J. Cell. Biol.* **91**, 113–125.

Sinanoğlu, O. (1981). *J. Chem. Phys.* **75**, 463–468; *Chem. Phys. Lett.* **81**, 188–190.

Singer, S. J., and Nicholson , G. L. (1972). *Science* **175**, 720–731.

Small, P. A. (1953). *J. Appl. Chem.* **3**, 71–80.

Smyth, C. P. (1955). "Dielectric Behaviour and Structure," McGraw-Hill, New York.

Snook, I. K., and van Megen, W. (1979). *J. Chem. Phys.* **70**, 3099–3105.

Snook, I. K., and van Megen, W. (1980). *J. Chem. Phys.* **72**, 2907–2913.

Snook, I. K., and van Megen, W. (1981). *J. Chem. Soc. Faraday Trans. 2* **77**, 181–190.

Sonntag, H., Unterberger, B., and Zimontkowski, S. (1979). *Colloid Polym. Sci.* **257**, 286–291.

Staehelin, L. A., and Arntzen, C. J. (1979). *In* "Chlorophyll Organization and Energy Transfer in Photosynthesis," Ciba Foundation Symp., vol. 61, pp. 147–175.

Stillinger, F. H., and Rahman, A. (1974). *J. Chem. Phys.* **60**, 1545–1557.

Tabor, D. (1977). *J. Colloid Interface Sci.* **58**, 2–13.

Tabor, D. (1982). *In* "Colloid Dispersions" (J. W. Goodwin, ed.), ch. 2, pp. 23–46. Royal Society of Chemistry, London.

Tabor, D., and Winterton, R. H. S. (1969). *Proc. Roy. Soc. Lond. A* **312**, 435–450.

Takano, K., and Hachisu, S. (1978). *J. Colloid Interface Sci.* **66**, 124–129.

Talmon, Y., Evans, D. F., and Ninham, B. W. (1983). *Science* **221**, 1047–1048.

Tanford, C. (1973, 1980). "The Hydrophobic Effect," Wiley, New York.

Thompson, D. W. (1968). "On growth and form," 2nd ed., vol. I, ch. 2, Cambridge Univ. Press, London and New York.

Tiddy, G. J. T. (1980). *Phys. Rep.* **57**, 1–46.

Torrie, G. M., and Valleau, J. P. (1979). *Chem. Phys. Lett.* **65**, 343–346.

Tucker, E. E., Lane, E. H., and Christian, S. D. (1981). *J. Solut. Chem.* **10**, 1–20.

Umeyama, H., and Morokuma, K. (1977). *J. Am. Chem. Soc.* **99**, 1316–1332.

Usui, S., and Yamasaki, T. (1969). *J. Colloid Interface Sci.* **29**, 629–638.

Usui, S., Yamasaki, T., and Shimolizak, J. (1967). *J. Phys. Chem.* **71**, 3195–3202.

van Blokland, P. H. G. M., and Overbeek, J. Th. G. (1978). *J. Chem. Soc. Faraday Trans. 1* **74**, 2637–2651;

van Blokland, P. H. G. M., and Overbeek, J. Th. G. (1979) *J. Colloid Interface Sci.* **68**, 96–100.

van Megen, W., and Snook, I. K. (1979). *J. Chem. Soc. Faraday Trans. 2* **75,** 1095–1102.

van Megen, W., and Snook, I. K. (1981). *J. Chem. Phys.* **74,** 1409–1411.

van Olphen, H. (1977). "An Introduction to Clay Colloid Chemistry," 2nd ed., ch. 10. Wiley, New York.

Van Oss, C. J., Absolom, D. R., and Neumann, A. W. (1980). *Colloids Surf.* **1,** 45–56.

Verwey, E. J. W., and Overbeek, J. Th. G. (1948). "Theory of the Stability of Lyophobic Colloids," Elsevier, Amsterdam.

Viani, B. E., Low, P. F., and Roth, C. B. (1983). *J. Colloid Interface Sci.* **96,** 229–244.

Visser, J. (1976). *Surf. Colloid Sci.* **8,** 3–79.

Von Hippel, A. R. (1958). "Dielectric materials and applications," Wiley, New York. See also "Handbook of Physics," 1958, ch. 7, McGraw-Hill, New York.

Vrij, A. (1976). *Pure Appl. Chem.* **48,** 471–483.

Wennerström, H., and Lindman, B. (1979). *Phys. Rep.* **52,** 1–86.

Wennerström, H., Jönsson, B., and Linse, P. (1982). *J. Chem. Phys.* **76,** 4665–4670.

Wesson, L. G. (1948). "Tables of Electric Dipole Moments," The Technology Press, MIT, Cambridge, Massachusetts.

Wieslander, Å, Christiansson, A., Rilfors, L., and Lindblom, G. (1980). *Biochemistry* **19,** 3650–3655.

Wilschut, J., Düzgünes, N., and Papahadjopoulos, D. (1981). *Biochemistry* **20,** 3126–3133.

Wobschall, D. (1971). *J. Colloid Interface Sci.* **36,** 385–396.

Zimmerberg, J., Cohen, F. S., and Finkelstein, A. (1980). *J. Gen. Physiol.* **75,** 251–270.

Zwanzig, R. (1963). *J. Chem. Phys.* **39,** 2251–2258.

Index